PARTIAL DEF

LIBRARY OF MATHEMATICS

edited by

WALTER LEDERMANN
D.Sc., Ph.D., F.R.S.Ed., Professor of
Mathematics, University of Sussex

PARTIAL DERIVATIVES

BY

P. J. HILTON

ROUTLEDGE & KEGAN PAUL
LONDON, HENLEY AND BOSTON

First published 1960
in Great Britain
by Routledge & Kegan Paul Ltd
39 Store Street
London. WC1E 7DD,
Broadway House, Newtown Road
Henley-on-Thames
Oxon. RG9 1EN and
9 Park Street
Boston, Mass. 02108, USA

Second impression 1963
Third impression 1965
Fourth impression 1969
Fifth impression 1973
Sixth impression 1978

ISBN 0 7100 4347 3

Printed in Great Britain by
Whitstable Litho Ltd, Whitstable, Kent

Preface

THIS book, like its predecessors in the same series, is intended primarily to serve the needs of the university student in the physical sciences. However, it begins where a really elementary treatment of the differential calculus (e.g., *Differential Calculus*,† in this series) leaves off. The study of physical phenomena inevitably leads to the consideration of functions of more than one variable and their rates of change; the same is also true of the study of statistics, economics, and sociology. The mathematical ideas involved are described in this book, and only the student familiar with the corresponding ideas for functions of a single variable should attempt to understand the extension of the method of the differential calculus to several variables.

The reader should also be warned that, with the deeper penetration into the subject which is required in studying functions of more than one variable, the mathematical arguments involved also take on a more sophisticated aspect. It should be emphasized that the *basic ideas* do not differ at all from those described in DC, but they are manipulated with greater dexterity in situations in which they are, perhaps, intuitively not so obvious. This remark may not console the reader bogged down in a difficult proof; but it may well happen (as so often in studying mathematics) that the reader will be given insight into the structure of a proof by following the examples provided and attempting the exercises. Thus the proofs of the theorems should not be the reader's prime concern on first reading.

P. J. HILTON

The University,
 Birmingham

† Referred to henceforth as DC.

Contents

vii

CONTENTS

CHAPTER ONE

Partial Derivatives and Partial Differentiation

(As mentioned in the Preface, this book is intended as a sequel to *Differential Calculus*, also published in this series. Throughout this book references to the earlier work will be cited as DC.

We recall that, if y is a given function of x, then in differentiating y with respect to x, we are measuring the rate of change of y with respect to x; in other words, we are measuring the way in which variations in x effect variations in y. Let us take an example. If a cylinder stands on a circular base of radius r and has height h, then its volume V is given by

$$V = \pi r^2 h \qquad (1.1)$$

We may then vary r and consider the consequent variations in V. The rate of change of V with r is given by

$$\frac{dV}{dr} = 2\pi r h \qquad (1.2)$$

On the other hand, we may regard r as fixed and consider the effect on V of varying h. We then have

$$\frac{dV}{dh} = \pi r^2 \qquad (1.3)$$

There are, of course, other variations one can consider: we may vary r and h simultaneously. But, for the time being, we will be content to consider the effect of varying *either r or h*. Then (1.2) expresses the rate of change of V with r when h is held constant and (1.3) expresses the rate of change of V with h when r is held constant. The essential fact in this example is that V is a function of two variables r and h which

1

are independent in the sense that variations in the value of one have no influence on the value of the other. To stress this question of independence consider a second example. A straight line AB of length 1 foot is divided at C, an equilateral triangle is erected on the base AC and a square on the base CB. If a is the total area, x the length of AC, and y the length of CB, then $a = \frac{\sqrt{3}}{4}x^2 + y^2$. Here a is a function of two variables x and y, but they are not independent since $x + y = 1$. Thus any variation of x implies a perfectly definite variation of y.

We now consider the general situation of which (1.1) is an example. We have a function $f(x, y)$ of two independent variables x, y and we write

$$z = f(x, y) \tag{1.4}$$

We digress for a moment to discuss a notational question. We may regard (1.4) as asserting that z is a function of the variables x, y so that z and f are the *same* function. Some authors prefer to express the fact that z is a function of x and y by writing

$$z = z(x, y)$$

instead of (1.4). This has the advantage (as we will see below) of avoiding duplication of notation, but it may prove awkward when we wish to consider two functions of x and y simultaneously. We will adhere to (1.4) but permit ourselves (as in DC) to regard z as a function (which, for example, we may differentiate) when convenient.

We now proceed to extend to functions of two variables the notions defined in DC Definitions 2.1, 2.2, and 2.3. The reader will observe that Definitions 1.1 and 1.2 are quite obvious generalizations of DC Definitions 2.1 and 2.2.

Definition 1.1. *We say that* f(x, y) *tends to the limit* 1 *as* (x, y) *tends to* (a, b) *if the difference between* f(x, y) *and* 1 *remains as small as we please so long as* (x, y) *remains sufficiently near to* (a, b), *while remaining distinct from* (a, b).

If $f(x, y)$ tends to l as (x, y) tends to (a, b) we write

$$\lim_{x, y \to a, b} f(x, y) = l$$

or

$$\lim_{x \to a, y \to b} f(x, y) = l;$$

but notice that in the second notation there is no significance in the order of the symbols $x \to a$ and $y \to b$.

Definition 1.2. *We say that* f(x, y) *is continuous at* (a, b) *if* $\lim_{x, y \to a, b}$ f(x, y) *exists and equals* f(a, b).

We make a few explanatory comments on these two definitions. The reader should notice particularly that the points that come into question are *all* those near to (a, b)—say, inside a circle of small radius with centre at (a, b)—and not merely those lying on the lines, $x=a$, $y=b$, parallel to the axes and passing through (a, b). We may, in fact, approach (a, b) by any route we please and the statement $\lim_{x, y \to a, b} f(x, y) = l$ means that the values of $f(x, y)$ along this route approach l as we approach (a, b). Thus we may immediately infer, by approaching (a, b) along the line $y=b$ or the line $x=a$ that, if $\lim_{x, y \to a, b} f(x, y) = l$, then

$$\lim_{x \to a} f(x, b) = l, \quad \lim_{y \to b} f(a, y) = l. \tag{1.5}$$

However the statement $\lim_{x, y \to a, b} f(x, y) = l$ is stronger than (1.5); that is, we can easily find a function $f(x, y)$ that approaches a certain limit l as x approaches a along $y=b$ or as y approaches b along $x=a$, but which does not approach l as (x, y) approaches (a, b) along some other route. A simple example is provided by the function

$$f(x, y) = \frac{xy}{x^2 + y^2}, \quad (x, y) \neq (0, 0); \tag{1.6}$$

3

we need not specify $f(0, 0)$ in discussing the limit of $f(x, y)$ as (x, y) approaches the origin. If (x, y) approaches the origin along $y=0$, $f(x, y)$ remains zero so that $\lim\limits_{x \to 0} f(x, 0)=0$ and, similarly, $\lim\limits_{y \to 0} f(0, y)=0$. On the other hand $f(x, y)=\frac{1}{2}$ on the line $x=y$ so that $f(x, y)$ tends to $\frac{1}{2}$ as (x, y) approaches the origin along $x=y$. Thus $\lim\limits_{x, y \to 0, 0} f(x, y)$ does not exist and so it is impossible to give $f(0, 0)$ a value so that $f(x, y)$ is continuous at the origin. If we do give $f(0, 0)$ the value zero, then, in an evident sense, $f(x, y)$ is continuous in each variable *separately* at the origin, but not in the two variables *simultaneously*.

Following this line of thought we proceed to define the notion that $f(x, y)$ is differentiable in each variable separately and postpone to Chapter Two the notion of differentiability in the two variables simultaneously. Thus we now consider the variables separately as in our example of the right-circular cylinder. Suppose then that we have $z=f(x, y)$; that is, the function f is to be regarded as defined over the (x, y)-plane. We may then hold y fixed and differentiate z with respect to x. We write the resulting derivative as $\dfrac{\partial z}{\partial x}$ or $\dfrac{\partial f}{\partial x}$ or z_x or f_x; if we particularly wish to stress that y is the variable held fixed† we may even write $(\partial z/\partial x)_y$ or $(\partial f/\partial x)_y$. Then $\dfrac{\partial z}{\partial x}$ is called the *partial derivative* of z with respect to x. Of course, the partial derivative of z with respect to y is defined similarly and a similar notation is used. The value of the partial derivative f_x at the point (a, b) is, naturally, written $f_x(a, b)$ and it is worth recalling explicitly its definition, according to DC Definition 2.3.

† See the remark following 1.7, for example.

Definition 1.3. *The partial differential coefficients at* $(x, y) =$ (a, b) *of the function* $f(x, y)$ *are*

$$f_x(a, b) = \lim_{h \to 0} \frac{f(a+h, b) - f(a, b)}{h},$$

$$f_y(a, b) = \lim_{k \to 0} \frac{f(a, b+k) - f(a, b)}{k}.$$

Notice that we distinguish (as in DC) between the derivative which is a *function*, and the differential coefficient which is a *number*. The phrase 'partial differential coefficient', which may not be familiar, seems to us useful, but we, and the reader, may sometimes prefer to speak of the 'partial derivative evaluated at (a, b)', or just 'partial derivative at (a, b)'.

We may sum up our description of partial differentiation in this way: partial differentiation is, in conception, no more subtle or obscure than 'ordinary' differentiation†—it is indeed nothing other than differentiation! The partial derivative with respect to x of $f(x, y)$ evaluated at (a, b) is just $g'(a)$, where $g(x)$ is the function of x defined by $g(x) = f(x, b)$. We reinforce this remark by stating immediately the following fundamental theorem.

Theorem 1.1. *If* $\frac{\partial f}{\partial x}$ *is identically zero near* (a, b), *then* f *depends only on* y *near* (a, b).

This result is a direct consequence of the theorem on functions of *one* variable that if the derivative vanishes identically the function is constant (Corollary 3.1 of DC).

We now give some examples of partial differentiation.

Example 1.1. Evaluate $\frac{\partial z}{\partial x}, \frac{\partial z}{\partial y}$ if $z = e^{kx} \cos ly$.

If y is held fixed, and we differentiate with respect to x,

† This point has been stressed because of the author's experience of students who can differentiate but can't 'do partials'!

5

we obtain $\dfrac{\partial z}{\partial x} = ke^{kx} \cos ly$. Similarly, if x is held fixed, and we differentiate with respect to y, we obtain $\dfrac{\partial z}{\partial y} = -le^{kx} \sin ly$.

Example 1.2. If (x, y) are Cartesian co-ordinates and (r, θ) are polar co-ordinates, then

$$\frac{\partial x}{\partial r} = \cos \theta, \ \frac{\partial y}{\partial r} = \sin \theta, \ \frac{\partial x}{\partial \theta} = -r \sin \theta, \ \frac{\partial y}{\partial \theta} = r \cos \theta;$$

$$\frac{\partial r}{\partial x} = \frac{x}{\surd(x^2 + y^2)}, \ \frac{\partial \theta}{\partial x} = \frac{-y}{x^2 + y^2}, \ \frac{\partial r}{\partial y} = \frac{y}{\surd(x^2 + y^2)}, \ \frac{\partial \theta}{\partial y} = \frac{x}{x^2 + y^2}.$$

Of course, the four equations in the second set are not valid at the origin. The equations, which are of fundamental importance, are easy consequences of the formulae $x = r \cos \theta$, $y = r \sin \theta$, $r = \surd(x^2 + y^2)$, $\theta = \tan^{-1} \dfrac{y}{x}$. We draw attention to one very striking feature of these equations. Substituting for x, y, in the formula for $\dfrac{\partial r}{\partial x}$, we find $\dfrac{\partial r}{\partial x} = \cos \theta$. Thus we have the unexpected result

$$\frac{\partial x}{\partial r} = \frac{\partial r}{\partial x}. \tag{1.7}$$

This appears at first sight to be in conflict with Theorem 2.5 of DC, which asserts that $\dfrac{dy}{dx} = 1/(dx/dy)$. However, if we use the more precise notation mentioned above, we write

$$\left(\frac{\partial x}{\partial r} \right)_\theta = \left(\frac{\partial r}{\partial x} \right)_y;$$

this notation shows that equation (1.7) has no connection with DC Theorem 2.5. On the other hand, DC Theorem 2.5 implies that $(\partial x/\partial r)_\theta = 1/(\partial r/\partial x)_\theta$, and the reader may verify that this is true—it follows immediately from $r = x \sec \theta$.

The importance of partial derivatives is that any variation in the pair of variables (x, y) may be regarded as composed of two variations, one in x and the other in y. Thus, if we wish to know how $f(x, y)$ varies as we pass from the point (a, b) to the point $(a+h, b+k)$, we may use the identity

$$f(a+h, b+k)-f(a, b)= \qquad (1.8)$$
$$(f(a+h, b+k)-f(a, b+k))+(f(a, b+k)-f(a, b))$$

to separate out the variation into a part due to a change in x from a to $a+h$ and a part due to a change in y from b to $b+k$. Thus the partial derivatives afford a measure of the total variation of the function; we revert to the significance of (1.8) in later chapters.

Just as for functions of one variable, we may define higher partial derivatives. However, the situation is somewhat more complicated owing to the possibility of mixed derivatives. Thus if $f(x, y)$ is a function of two variables, we may form four derivatives of the second order: in addition to

$$\frac{\partial}{\partial x}\left(\frac{\partial f}{\partial x}\right) \text{ and } \frac{\partial}{\partial y}\left(\frac{\partial f}{\partial y}\right)$$

there are the mixed derivatives

$$\frac{\partial}{\partial x}\left(\frac{\partial f}{\partial y}\right) \text{ and } \frac{\partial}{\partial y}\left(\frac{\partial f}{\partial x}\right).$$

We write these as $\dfrac{\partial^2 f}{\partial x^2}, \dfrac{\partial^2 f}{\partial y^2}, \dfrac{\partial^2 f}{\partial x \partial y}, \dfrac{\partial^2 f}{\partial y \partial x}$, or $f_{xx}, f_{yy}, f_{xy}, f_{yx}$; there are similar notations for derivatives of higher order. Fortunately the situation is simplified by the fact that, in fairly general circumstances,

$$\frac{\partial^2 f}{\partial x \partial y}=\frac{\partial^2 f}{\partial y \partial x}; \qquad (1.9)$$

the equality holds for example if each expression in (1.9) exists and is continuous. We will henceforth assume (1.9), but will give the proof in an appendix. For the time being we are content if the reader realizes that (1.9) is not an *immediate* consequence of definition.

Finally we remark that what has been said for functions of two variables extends in an obvious way to functions of more than two variables. In fact in the technical developments we introduce later we will frequently be concerned with functions of at least three variables. However, when the extension is quite obvious we will state our definitions and theorems only for functions of two variables.

Exercises

1. Find $f_x, f_y, f_{xx}, f_{xy}, f_{yx}, f_{yy}$ if $f(x, y) = \sin(x-y)e^{x+y}$.

2. Find f_x, f_y, f_z if $f(x, y, z) = \left(\dfrac{x}{y} + \dfrac{y}{z} - \dfrac{z}{x}\right)\log\left(\dfrac{x}{y} - \dfrac{y}{z} + \dfrac{z}{x}\right)$.

3. If $u = \sqrt{(x^2+y^2)} + x$, $v = \sqrt{(x^2+y^2)} - x$ find u_x, v_x, u_y, v_y. Solve for x, y in terms of u, v and find x_u, x_v, y_u, y_v.

4. If $z = \dfrac{1}{\sqrt{(x^2 - 2xy + a^2)}}$, show that $\dfrac{\partial}{\partial x}\left(x^2\dfrac{\partial z}{\partial x}\right) = \dfrac{\partial}{\partial y}\left((y^2 - a^2)\dfrac{\partial z}{\partial y}\right)$.

5. Evaluate $f_x(0, 0)$, $f_y(0, 0)$ where $f = \dfrac{x^3 + 2y^3}{x^2 + y^2}$.

CHAPTER TWO

Differentiability and Change of Variables

1. DIFFERENTIABILITY

The fundamental rules given in Chapter Two of DC for differentiating sums, products, and quotients of functions apply, of course, equally well to partial differentiation. However, the most important rule in partial differentiation is the generalization of Theorem 2.5 of DC on change of variables.

If we consult the proof of Theorem 2.5 of DC, we see that the essential idea was that, if $y=f(x)$, then $f'(a)=c$ *if and only if* $f(a+h)=f(a)+h(c+\epsilon(h))$, where the function $\epsilon(h)\to 0$ as $h\to 0$. It is this idea which we will generalize for the purpose of discussing the formula for changing variables in partial derivatives. The actual discussion of the formula will occupy us in the next section. In this section we will consider the generalization to functions of two variables of the relation

$$f(a+h)=f(a)+h(f'(a)+\epsilon(h)), \qquad (2.1)$$

where $\epsilon(h)\to 0$ as $h\to 0$. Now let $f(x, y)$ be a function of two variables and let

$$f(a+h, b+k)=f(a, b)+h(c+\epsilon(h, k))+k(d+\eta(h, k)), \qquad (2.2)$$

where $\epsilon(h, k)\to 0$, $\eta(h, k)\to 0$ as $h, k\to 0$. Then, putting $k=0$ and letting $h\to 0$, we immediately infer that $c=f_x(a, b)$, and similarly $d=f_y(a, b)$. However the converse is not always true; that is to say, even when $f_x(a, b)$, $f_y(a, b)$ both exist it is *not* always true that

$$f(a+h,b+k)=f(a,b)+h(f_x(a,b)+\epsilon(h,k))+k(f_y(a,b)+\eta(h,k)) \qquad (2.3)$$

where $\epsilon(h, k)\to 0$, $\eta(h, k)\to 0$ as $h, k\to 0$. We show this by an example; in fact, we use the example invoked in discussing Definitions 4.1 and 4.2. Thus let $f(x, y)$ be defined by

$$f(x, y) = \frac{xy}{x^2+y^2}, \ (x, y) \neq (0, 0), f(0, 0) = 0.$$

Then $f(x, 0) = 0 = f(0, y)$. Thus $f_x(0, 0) = 0 = f_y(0, 0)$. If (2.3) held with $a = 0, b = 0$, we would have $f(h, k) = \frac{hk}{h^2+k^2} = h\epsilon(h, k) + k\eta(h\ k)$. Let us approach $(0, 0)$ along $x = y$, i.e., put $h = k$. Then $\frac{1}{2} = h(\epsilon(h, h) + \eta(h, h))$. Letting $h \to 0$, we have $\frac{1}{2} = 0$, so that (2.3) does not hold for this particular function $f(x, y)$ although $f_x(0, 0)$ and $f_y(0, 0)$ are defined.

However we may ensure (2.3) by a mild restriction.

Theorem 2.1. *If* $f_x(x, y)$ *is continuous at* (a, b), *then* (2.3) *holds*.

Consider (1.8). By the Mean Value Theorem for functions of one variable $f(a+h, b+k) - f(a, b+k) = hf_x(a+\theta h, b+k)$, for some θ satisfying $0 < \theta < 1$. By the continuity of $f_x(x, y)$, $f_x(a+\theta h, b+k) = f_x(a, b) + \epsilon(h, k)$, where $\epsilon(h, k) \to 0$ as $h, k \to 0$. Applying (2.1) to the differential coefficient $f_y(a, b)$, we have $f(a, b+k) - f(a, b) = k(f_y(a, b) + \eta(k))$, where $\eta(k) \to 0$ as $k \to 0$. Putting these facts together, we get (2.3)—the fact that η does not depend on h does not disturb us!

Definition 2.1. *We say that* $f(x, y)$ *is differentiable at* (a, b) *if* (2.3) *holds. We say that* $f(x, y)$ *is continuously differentiable at* (a, b) *if the partial derivatives* $f_x(x, y)$, $f_y(x, y)$ *are continuous at* (a, b).

Notice that in the light of Theorem 2.1 a continuously differentiable function *is* differentiable, so our terminology is justified! Since we will not concern ourselves very much with badly-behaved† functions, the reader may assume, unless the contrary is stated, that the functions entering our discussions are continuously differentiable.

† One must be aware of the possibility of bad behaviour but only the specialist in 'behaviour studies' encourages it!

Relation (2.3), while clearly generalizing (2.1), may seem to the reader a very obscure criterion of differentiability. It may be made more acceptable by writing ϵ, η for $\epsilon(h, k)$, $\eta(h, k)$, so long as it is remembered that ϵ and η each depend on h and k. Then (2.3) becomes

$$f(a+h, b+k)=f(a, b)+hf_x(a, b)+kf_y(a, b)+h\epsilon+k\eta,$$
(2.4)

where ϵ, $\eta \to 0$ as h, $k \to 0$. If we write ρ for $\sqrt{(h^2+k^2)}$ we may express the criterion for differentiability in the even neater form

$$f(a+h, b+k)=f(a, b)+hf_x(a, b)+kf_y(a, b)+\rho\epsilon,$$
(2.5)

where $\epsilon \to 0$ as $\rho \to 0$.

It is immediately obvious from (2.5) or its variants that $f(x, y)$ is continuous at (a, b) if it is differentiable; this generalizes the corresponding statement for functions of one variable. However, our faithful example of $\dfrac{xy}{x^2+y^2}$ shows that $f_x(a, b)$ and $f_y(a, b)$ may both exist without $f(x, y)$ being continuous at (a, b).

We work one example before returning to the problem of change of variables.

Example 2.1. If $f(x, y)=\dfrac{x^3y^3}{x^2+y^2}$, $(x, y) \neq (0, 0)$, and $f(0, 0)=0$, then $f(x, y)$ is differentiable at the origin. For we show first that $f_x(0, 0)=0$, $f_y(0, 0)=0$ so we must prove that

$$\frac{h^3k^3}{h^2+k^2}=\sqrt{(h^2+k^2)} \cdot \epsilon, \text{ where } \epsilon \to 0 \text{ as } h, k \to 0;$$ or, simply

that $\dfrac{h^3k^3}{(h^2+k^2)^{3/2}} \to 0$ as h, $k \to 0$. Now, since $k^2 \geqslant 0$,

$\left| \dfrac{h^3}{(h^2+k^2)^{3/2}} \right| \leqslant 1$, so that $\left| \dfrac{h^3k^3}{(h^2+k^2)^{3/2}} \right| \leqslant |k|^3$, and the conclusion follows.

11

2. CHANGE OF VARIABLES IN PARTIAL DIFFERENTIATION

We now come to the promised generalization of Theorem 2.5 of DC.

Theorem 2.2. *If z is a differentiable function of* (u, v) *and* u, v *are continuous functions of* x, y, *with partial derivatives, then*

$$\frac{\partial z}{\partial x} = \frac{\partial z}{\partial u}\frac{\partial u}{\partial x} + \frac{\partial z}{\partial v}\frac{\partial v}{\partial x}, \quad \frac{\partial z}{\partial y} = \frac{\partial z}{\partial u}\frac{\partial u}{\partial y} + \frac{\partial z}{\partial v}\frac{\partial v}{\partial y}.$$

It is obviously sufficient to prove† the formula for $\frac{\partial z}{\partial x}$. The idea of the proof is exactly as for Theorem 2.5 of DC. Let $z = F(u, v)$, $u = f(x, y)$, $v = g(x, y)$, $u + p = f(x+h, y)$, $v + q = g(x+h, y)$. Then, by definition,

$$\frac{\partial z}{\partial x} = \lim_{h \to 0} \frac{F(u+p, v+q) - F(u, v)}{h}.$$

Now, by (2.3), $F(u+p, v+q) - F(u, v) = p(F_u(u, v) + \epsilon) + q(F_v(u, v) + \eta)$, where $\epsilon, \eta \to 0$ as $p, q \to 0$. But, by (2.1), $p = f(x+h, y) - f(x, y) = h(f_x(x, y) + \epsilon')$, where $\epsilon' \to 0$ as $h \to 0$, and $q = g(x+h, y) - g(x, y) = h(g_x(x, y) + \eta')$, where $\eta' \to 0$ as $h \to 0$. Moreover, $p, q \to 0$ as $h \to 0$, since f, g are continuous, so that $\epsilon, \eta \to 0$ as $h \to 0$. Thus

$$F(u+p, v+q) - F(u, v) = h\{(f_x(x, y) + \epsilon')(F_u(u, v) + \epsilon) + (g_x(x, y) + \eta')(F_v(u, v) + \eta)\},$$

whence $\dfrac{F(u+p, v+q) - F(u, v)}{h} = f_x(x, y)F_u(u, v) +$

$g_x(x, y)F_v(u, v) + \sigma$, where $\sigma = \epsilon f_x(x, y) + \epsilon' F_u(u, v) + \epsilon\epsilon' + \eta g_x(x, y) + \eta' F_v(u, v) + \eta\eta'$, so that $\sigma \to 0$ as $h \to 0$. Thus

$$\frac{\partial z}{\partial x} = f_x(x, y)F_u(u, v) + g_x(x, y)F_v(u, v) = \frac{\partial z}{\partial u}\frac{\partial u}{\partial x} + \frac{\partial z}{\partial v}\frac{\partial v}{\partial x},$$

and the theorem is proved. The reader should note that this theorem *does* generalize Theorem 2.5 of DC. For if z really

† We repeat what we said in DC—the reader should try to understand proofs but should not be required to memorize them.

only depends on u, and u is a function of x only, then $\dfrac{\partial z}{\partial u} = \dfrac{dz}{du}$, $\dfrac{\partial u}{\partial x} = \dfrac{du}{dx}$, $\dfrac{\partial z}{\partial v} = 0$, $\dfrac{\partial z}{\partial x} = \dfrac{dz}{dx}$, and we have $\dfrac{dz}{dx} = \dfrac{dz}{du}\dfrac{du}{dx}$.

The reader should not worry too much about the precise conditions under which the theorem holds; suffice it that it is true if z, u, and v are differentiable functions.

Example 2.2. Let $x = r\cos\theta$, $y = r\sin\theta$; then

$$\frac{\partial z}{\partial r} = \frac{\partial z}{\partial x}\cdot\frac{\partial x}{\partial r} + \frac{\partial z}{\partial y}\frac{\partial y}{\partial r} = \cos\theta\frac{\partial z}{\partial x} + \sin\theta\frac{\partial z}{\partial y}, \text{ and } \frac{\partial z}{\partial\theta} = \frac{\partial z}{\partial x}\frac{\partial x}{\partial\theta} + \frac{\partial z}{\partial y}\frac{\partial y}{\partial\theta}$$

$$= -r\sin\theta\frac{\partial z}{\partial x} + r\cos\theta\frac{\partial z}{\partial y}. \text{ For example, if } z = xy, \text{ then } \frac{\partial z}{\partial r} =$$

$\cos\theta\cdot y + \sin\theta\cdot x = 2r\sin\theta\cos\theta$ and $\dfrac{\partial z}{\partial\theta} = -r\sin\theta\cdot y + r\cos\theta\cdot x = r^2(\cos^2\theta - \sin^2\theta)$. These formulae may be verified directly from $z = r^2\sin\theta\cos\theta$.

Example 2.3. Theorem 2.2 generalizes the sum, product, and quotient rules. For example, let $z = \dfrac{f(x)}{g(x)}$; put $u = f(x)$, $v = g(x)$. Replacing $\dfrac{\partial z}{\partial x}, \dfrac{\partial u}{\partial x}, \dfrac{\partial v}{\partial x}$, by $\dfrac{dz}{dx}, \dfrac{du}{dx}, \dfrac{dv}{dx}$, as we may since z, u, v only depend on the variable x, we have $z = \dfrac{u}{v}$, and $\dfrac{dz}{dx} = \dfrac{\partial z}{\partial u}\dfrac{du}{dx} + \dfrac{\partial z}{\partial v}\dfrac{dv}{dx} = \dfrac{1}{v}f'(x) - \dfrac{u}{v^2}g'(x) = \dfrac{f'(x)}{g(x)} - \dfrac{f(x)g'(x)}{(g(x))^2}$

$$= \frac{g(x)f'(x) - f(x)g'(x)}{(g(x))^2}.$$

In this last example, we have invoked the special case in which u and v are functions of x alone, so that

$$\frac{dz}{dx} = \frac{\partial z}{\partial u}\frac{du}{dx} + \frac{\partial z}{\partial v}\frac{dv}{dx}. \tag{2.6}$$

In particular we could consider $z=f(x, y)$ where y is itself a function of x. Then $u=x$, $v=y$ so that

$$\frac{dz}{dx}=\frac{\partial z}{\partial x}+\frac{\partial z}{\partial y}\frac{dy}{dx} \tag{2.7}$$

In this equation, $\frac{dz}{dx}$ has to be interpreted as the derivative of z regarded as a function of the single variable x and $\frac{\partial z}{\partial x}$ is the partial derivative of z *regarded as a function of x and y.*

Example 2.4. Find $\frac{dz}{dx}$ if $z=x\sin(x+y)$, where $y=\sin^{-1}x$.

Now $\frac{\partial z}{\partial x}=\sin(x+y)+x\cos(x+y)$, $\frac{\partial z}{\partial y}=x\cos(x+y)$,

$\frac{dy}{dx}=\frac{1}{\sqrt{(1-x^2)}}$. Thus

$$\frac{dz}{dx}=\sin(x+y)+x\cos(x+y)+\frac{x}{\sqrt{(1-x^2)}}\cos(x+y).$$

The reader will note that formulae (2.6) and (2.7) give expressions for differentiating functions of *one* variable. Thus our work in partial differentiation has led us to useful conclusions in the 'ordinary' differential calculus.

We may also apply the rule for changing variables given by Theorem 2.2 to higher-order partial derivatives. We demonstrate by an example.

Example 2.5. Functions $f(x, y)$ satisfying

$$\frac{\partial^2 f}{\partial x^2}+\frac{\partial^2 f}{\partial y^2}=0 \tag{2.8}$$

are called harmonic functions and play a vital role in potential theory. It is thus important to know that if z is a function of x, y and hence also of r, θ, then

$$\frac{\partial^2 z}{\partial x^2}+\frac{\partial^2 z}{\partial y^2}=\frac{\partial^2 z}{\partial r^2}+\frac{1}{r}\frac{\partial z}{\partial r}+\frac{1}{r^2}\frac{\partial^2 z}{\partial \theta^2}. \tag{2.9}$$

Before proving this we make a general remark about technique. If u, v are given as functions of independent variables p, q, say, and we wish to prove equality between a formula involving derivatives with respect to u, v and one involving derivatives with respect to p, q then we should start with the second formula and use Theorem 2.2 to transform to the first. Thus, in this example, we have x, y as functions of r, θ and so we start from the right-hand side of (2.9). Of course, in this case, we can easily obtain r, θ as functions of x, y but such a procedure may, in general, be extremely inconvenient. Thus to exemplify the point of technique under discussion, we will consider the expression

$$\frac{\partial^2 z}{\partial r^2} + \frac{1}{r} \frac{\partial z}{\partial r} + \frac{1}{r^2} \frac{\partial^2 z}{\partial \theta^2}.$$

Now $\dfrac{\partial z}{\partial r} = \dfrac{\partial z}{\partial x} \dfrac{\partial x}{\partial r} + \dfrac{\partial z}{\partial y} \dfrac{\partial y}{\partial r} = \cos\theta \dfrac{\partial z}{\partial x} + \sin\theta \dfrac{\partial z}{\partial y}$; similarly,

$$\frac{\partial z}{\partial \theta} = -r \sin\theta \frac{\partial z}{\partial x} + r \cos\theta \frac{\partial z}{\partial y}.$$

We may write these two equations in the operational form

$$\frac{\partial}{\partial r} = \cos\theta \frac{\partial}{\partial x} + \sin\theta \frac{\partial}{\partial y}, \quad \frac{\partial}{\partial \theta} = -r \sin\theta \frac{\partial}{\partial x} + r \cos\theta \frac{\partial}{\partial y}.$$

These equations mean that the *differential operators* $\dfrac{\partial}{\partial r}$ and $\cos\theta \dfrac{\partial}{\partial x} + \sin\theta \dfrac{\partial}{\partial y}$, and likewise $\dfrac{\partial}{\partial \theta}$ and $-r \sin\theta \dfrac{\partial}{\partial x} + r \cos\theta \dfrac{\partial}{\partial y}$, have the same effect when applied to a differentiable function. In particular this is true for the functions $\dfrac{\partial z}{\partial x}$ and $\dfrac{\partial z}{\partial y}$, which we assume differentiable. Thus

$$\frac{\partial^2 z}{\partial^2 r} = \frac{\partial}{\partial r}\left(\frac{\partial z}{\partial r}\right)$$
$$= \frac{\partial}{\partial r}\left(\cos\theta \frac{\partial z}{\partial x} + \sin\theta \frac{\partial z}{\partial y}\right)$$

15

$$=\cos\theta\frac{\partial}{\partial r}\left(\frac{\partial z}{\partial x}\right)+\sin\theta\frac{\partial}{\partial r}\left(\frac{\partial z}{\partial y}\right)$$

$$=\cos\theta\left(\cos\theta\frac{\partial}{\partial x}+\sin\theta\frac{\partial}{\partial y}\right)\frac{\partial z}{\partial x}+\sin\theta\left(\cos\theta\frac{\partial}{\partial x}+\sin\theta\frac{\partial}{\partial y}\right)\frac{\partial z}{\partial y}.$$

$$=\cos^2\theta\frac{\partial^2 z}{\partial x^2}+2\sin\theta\cos\theta\frac{\partial^2 z}{\partial x\partial y}+\sin^2\theta\frac{\partial^2 z}{\partial y^2}.$$

Similarly

$$\frac{\partial^2 z}{\partial\theta^2}=\frac{\partial}{\partial\theta}\left(-r\sin\theta\frac{\partial z}{\partial x}+r\cos\theta\frac{\partial z}{\partial y}\right)$$

$$=-r\cos\theta\frac{\partial z}{\partial x}-r\sin\theta\frac{\partial}{\partial\theta}\frac{\partial z}{\partial x}-r\sin\theta\frac{\partial z}{\partial y}+r\cos\theta\frac{\partial}{\partial\theta}\frac{\partial z}{\partial y}$$

$$=-r\frac{\partial z}{\partial r}-r\sin\theta\left(-r\sin\theta\frac{\partial}{\partial x}+r\cos\theta\frac{\partial}{\partial y}\right)\frac{\partial z}{\partial x}$$

$$+r\cos\theta\left(-r\sin\theta\frac{\partial}{\partial x}+r\cos\theta\frac{\partial}{\partial y}\right)\frac{\partial z}{\partial y}$$

$$=-r\frac{\partial z}{\partial r}+r^2\sin^2\theta\frac{\partial^2 z}{\partial x^2}-2r^2\sin\theta\cos\theta\frac{\partial^2 z}{\partial x\partial y}+r^2\cos^2\theta\frac{\partial^2 z}{\partial y^2}.$$

It immediately follows that

$$\frac{\partial^2 z}{\partial r^2}+\frac{1}{r}\frac{\partial z}{\partial r}+\frac{1}{r^2}\frac{\partial^2 z}{\partial\theta^2}=\frac{\partial^2 z}{\partial x^2}+\frac{\partial^2 z}{\partial y^2},$$

and (2.9) is proved.

3. DIFFERENTIALS

We begin this section with a definition and discussion of differentials for functions of a single variable. We may re-write (2.1) as

$$f(a+h)-f(a)=hf'(a)+h\,\epsilon(h),$$

where $\epsilon(h)\to 0$ as $h\to 0$. Writing δf for $f(a+h)-f(a)$ and df for $hf'(a)$, we get

$$\delta f=df+h\,\epsilon(h).$$

We call df the *differential* of f at $x=a$. Notice that it is a multiple of h, the coefficient being $f'(a)$; this serves to explain the use of the term 'differential coefficient' for $f'(a)$.

Now consider in particular the function $f(x)=x$. Strictly we should make a distinction in our notation between the variable x and this function, but if we do not make this distinction then we get

$$dx=h,$$

since $f'(a)=1$ for this function f. Thus we may write

$$df=f'(a)dx. \qquad (2.10)$$

We emphasize that (2.10) merely expresses the fact that the differentials of the functions f and x stand in a certain constant ratio; there has been no mention of 'infinitesimal quantities' or of δf and δx becoming df and dx 'in the limit'.

On the other hand we may regard (2.10) as describing approximately, for small changes in x, the consequent change in $f(x)$; the error is $\delta f - df = h\epsilon(h)$, which may, of course, be large if h is large. Indeed the error is, in geometrical language, just that due to replacing the curve $y=f(x)$ by its tangent at $x=a$ (see Fig. 1).

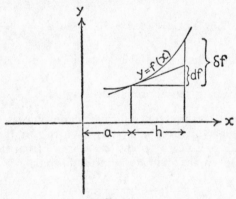

Fig. 1

For example if $f(x)=x^3$ and $a=1$, then if $h(=dx)=1$, $df=3$, while the actual change in f is given by $\delta f=2^3-1=7$. But if $h=1/10$ then $df=\cdot3$ and $\delta f=\cdot331$.

What has been said for functions of a single variable generalizes to functions of two (or more) variables. Replacing (2.1) by (2.5) we have

$$f(a+h, b+k)-f(a, b)=hf_x(a, b)+kf_y(a, b)+\rho\epsilon,$$

or

$$\delta f=df+\rho\epsilon,$$

where $\delta f=f(a+h, b+k)-f(a, b)$, $df=hf_x(a, b)+kf_y(a, b)$. In particular if f is the *co-ordinate function* $f(x, y)=x$, then $f_x(a, b)=1$, $f_y(a, b)=0$, and if $f(x, y)=y$ then $f_x(a, b)=0$, $f_y(a, b)=1$. Thus $dx=h$, $dy=k$ and

$$df=f_x(a, b)dx+f_y(a, b)dy. \tag{2.11}$$

Here the differential df is expressed as a linear combination of the differentials dx and dy; and we may use (2.11) to estimate the change in f resulting from small changes in x and y, the accuracy of the estimate depending, of course, on the size of the neglected term $\delta f-df=\rho\epsilon$; recall that $\rho=\sqrt{(h^2+k^2)}=\sqrt{(dx)^2+(dy)^2}$ and $\epsilon\to0$ as $\rho\to0$.

Example 2.6. Let $z=\sqrt{(x^2+y^2)}$. Then $dz=\dfrac{x}{\sqrt{(x^2+y^2)}}dx+\dfrac{y}{\sqrt{(x^2+y^2)}}dy$. Thus if we change (a, b) to $(a+h, b+k)$, the change in z is given approximately by $dz=\dfrac{ah+bk}{\sqrt{(a^2+b^2)}}$. The geometrical interpretation of this relationship is that, if s is the distance of the point (a, b) from the origin and s_1 is the distance of a nearby point $(a+h, b+k)$ from the origin, then s_1 is given approximately by the formula

$$s_1=s+\frac{ah+bk}{s}.$$

We have spoken of (2.11) giving us an estimate of the change in f consequent on small changes in x and y. Now it may happen that in practice we are only able to estimate the values of x and y with a certain known accuracy; we may then use (2.11) to estimate the accuracy of our value for f. Reverting to example 2.5, suppose that we measure x and y to be equal to a and b respectively with errors not exceeding p per cent, q per cent respectively. Then $|dx| < \dfrac{pa}{100}$, $|dy| < \dfrac{qb}{100}$, so that

$$|dz| < \frac{pa^2 + qb^2}{100\sqrt{(a^2 + b^2)}}.$$

Thus the percentage error in our value $\sqrt{(a^2 + b^2)}$ for z should not exceed r per cent, where $r = \dfrac{pa^2 + qb^2}{a^2 + b^2}$. For example, if $p = q$, then $p = q = r$, so that the error in z is approximately the same as that in x and y, provided, of course, the errors are small.

On the other hand, a quite different interpretation of equation (2.11) may be given. Let $z = f(x, y)$ be a differentiable function and let x, y be differentiable functions of a parameter t. Then z is a function of t and

$$\frac{dz}{dt} = \frac{\partial z}{\partial x}\frac{dx}{dt} + \frac{\partial z}{\partial y}\frac{dy}{dt} = f_x\frac{dx}{dt} + f_y\frac{dy}{dt}. \qquad (2.12)$$

Thus (2.11) may be interpreted as giving the rate of change of z with t in terms of the rates of change of x and y with t for any parameter t.

If we think of $z = f(x, y)$ as the equation describing a surface in space, then $x = x(t)$, $y = y(t)$ are equations of a curve in this surface. If $x(t_0) = a$, $y(t_0) = b$, $f(a, b) = c$, then it may be shown that the tangent to this curve at the point with parameter t_0 is

$$\frac{x - a}{dx/dt} = \frac{y - b}{dy/dt} = \frac{z - c}{dz/dt}.$$

In the light of (2.12) this line lies in the plane

$$z - c = (x - a)f_x(a, b) + (y - b)f_y(a, b). \qquad (2.13)$$

This plane, which we have seen contains the tangents at (a, b, c) to any curve in the surface passing through (a, b, c), is called the *tangent plane* to the surface at (a, b, c). Now by (2.5) $f(x, y) - c = (x - a)f_x(a, b) + (y - b)f_y(a, b) + \rho\epsilon$, where

19

$$\epsilon \to 0 \text{ as } \rho = \sqrt{(x-a)^2 + (y-b)^2} \to 0.$$

Thus we may regard the function $z = c + (x-a)f_x(a, b) + (y-b) f_y(a, b)$ as a linear approximation to $f(x, y)$ near (a, b) and we see that the function approximating linearly to $f(x, y)$ at (a, b) determines precisely the tangent plane to the surface $z = f(x, y)$ at the point (a, b, c) where $c = f(a, b)$. This is entirely analogous to the situation for functions of a single variable described earlier (see also DC, p. 35).

Notice that the line

$$\frac{x-a}{f_x(a, b)} = \frac{y-b}{f_y(a, b)} = \frac{z-c}{-1} \tag{2.14}$$

passes through the point (a, b, c) and is normal to the plane (2.13). The line (2.14) is called the *normal* to the surface $z = f(x, y)$ at the point (a, b, c).

In any interpretation of (2.11) the question arises: if $p(x, y)$, $q(x, y)$ are functions of x, y, is $pdx + qdy$ a differential? This question is equivalent to the question: does there exist a function $f(x, y)$ such that $f_x = p$, $f_y = q$? The answer is provided by:

Theorem 2.3. *A function* f(x, y) *exists such that* $f_x = p$, $f_y = q$, *if and only if* $p_y = q_x$.

Here we assume the usual differentiability conditions. Thus if $f_x = p$, $f_y = q$, then

$$p_y = f_{yx} = f_{xy} = q_x.$$

Conversely, suppose $p_y = q_x$ and let $g(x, y)$ be any function† such that $g_x = p$. Then

$$g_{xy} = g_{yx} = p_y = q_x;$$

in other words $\dfrac{\partial}{\partial x}(g_y - q) = 0$. This means that $g_y - q$ is a function of y alone, say $g_y - q = h$. Let $k(y)$ be such that $k'(y) = h(y)$. Then put $f(x, y) = g(x, y) - k(y)$. We have $f_x =$

---

† We assume that continuous functions can be integrated, and thus appear as derivatives of functions. This is a fundamental theorem of the integral calculus.

$g_x = p$ and $f_y = g_y - k' = g_y - h = q$, so that the theorem is proved.

Example 2.7. Prove that $(3x^2 + 8xy + 6y^2)dx + (4x^2 + 12xy - 6y^2)dy$ is a differential and find a function f of which it is the differential. Here $p = 3x^2 + 8xy + 6y^2$, $q = 4x^2 + 12xy - 6y^2$. Then $p_y = 8x + 12y = q_x$, so that $p dx + q dy$ is a differential.

Put $g = x^3 + 4x^2 y + 6xy^2$ so that $g_x = p$; then $g_y = 4x^2 + 12xy$, $g_y - q = 6y^2$, a function of y only as theory predicted. Then if $h(y) = 6y^2$, we may take $k(y) = 2y^3$ so that $k'(y) = h(y)$ and we may take $f(x, y) = g(x, y) - k(y) = x^3 + 4x^2 y + 6xy^2 - 2y^3$.

Exercises.

1. Prove that $(xy)^{1/3}$ is not differentiable at the origin.

2. Generalize formula (2.3) to functions of three variables and show that $f(x, y, z)$ is differentiable at (a, b, c) if two of the partial derivatives f_x, f_y, f_z are continuous at (a, b, c).

3. By transferring to cylindrical polar coordinates, or otherwise, show that $\dfrac{\partial^2 V}{\partial x^2} + \dfrac{\partial^2 V}{\partial y^2} + \dfrac{\partial^2 V}{\partial z^2} = 0$, if $V = \tan^{-1}\dfrac{y}{x}$ or $V = z \tan^{-1}\dfrac{y}{x}$ or $V = \dfrac{\tan^{-1} y/x}{\sqrt{(x^2 + y^2 + z^2)}}$.

4. If $\phi(u, v)$ becomes $f(x, y)$ under the substitution $u = u(x, y)$, $v = v(x, y)$, prove that
$$f_{xx} = \phi_u u_{xx} + \phi_v v_{xx} + \phi_{uu} u_x^2 + 2\phi_{uv} u_x v_x + \phi_{vv} v_x^2.$$
If $u = \dfrac{x}{x^2 + y^2}$, $v = \dfrac{y}{x^2 + y^2}$, prove that $\phi_{uu} + \phi_{vv} = (x^2 + y^2)^2 (f_{xx} + f_{yy})$.

5. If $x = e^u \cos v$, $y = e^u \sin v$, prove that, with the notation of the previous question,
$$\phi_{uu} + \phi_{vv} = (x^2 + y^2)(f_{xx} + f_{yy}).$$

6. Let $f(x, y, z)$ become $\phi(u, v)$ on the surface $x = a(u + v)$, $y = b(u - v)$, $z = uv$; prove that

$$2(\phi_{uu}+\phi_{vv})=\left(\frac{x^2}{a^2}+\frac{y^2}{b^2}\right)f_{zz}+4(a^2f_{xx}+b^2f_{yy}+xf_{xz}-yf_{yz}).$$

7. If $z=f(g(x)+h(y))$, prove that $pq(qu-pv)=s(q^2r-p^2t)$, where $p=z_x$, $q=z_y$, $r=z_{xx}$, $s=z_{xy}$, $t=z_{yy}$, $u=z_{xxy}$, $v=z_{xyy}$.

8. Generalize Theorem 2.2 to functions of three variables.

9. The area of a triangle ABC is computed from the formula $\varDelta=\frac{1}{2}bc \sin A$. Estimate the percentage error in $\varDelta$ if b, c, A are subject to errors of p per cent, q per cent, r per cent.

10. The length, c, of a side of a triangle ABC is computed from the formula $c^2=a^2+b^2-2ab \cos C$. Estimate the percentage error in c if a, b, C are subject to errors of p per cent, q per cent, r per cent. (The reader may find it instructive in this and the previous example to give numerical values to the quantities involved.)

11. Find the tangent plane and normal to the cone $z=x^2+y^2$ at the point $(1, 2, 5)$.

12. Generalize Theorem 2.3 to functions of three variables by showing that a function $f(x, y, z)$ exists with $f_x=p$, $f_y=q$, $f_z=r$ if and only if

$$p_y=q_x, \ p_z=r_x, \ q_z=r_y.$$

Verify the theorem if $p=(x+1)e^{x+y}\sin z$, $q=xe^{x+y}\sin z$, $r=xe^{x+y}\cos z$.

CHAPTER THREE
Implicit Functions

1. FUNDAMENTAL THEOREM

It frequently happens that the relationship between two variables x and y is not expressed by an equation of the type $y=f(x)$ but by an equation of the type $f(x, y)=0$; a similar remark holds for three or more variables. Thus, for example, the circle centre the origin, radius a, is conveniently described by the equation $x^2+y^2-a^2=0$; the general conic section by the equation $ax^2+2hxy+by^2+2gx+2fy+c=0$. It would be inconvenient in either case to express y *explicitly* as a function of x and far more convenient to deal with the *implicit* relationship indicated by the equation $f(x, y)=0$. Sometimes it is not only inconvenient but even impossible to 'solve' the equation $f(x, y)=0$ for y in any effective sense (consider the equation $xy=\cos(x+y)$) and consequently it is important, first, to know under what conditions the relation $f(x, y)=0$ does determine y as a function of x and, second, to be able to calculate $\frac{dy}{dx}$ at any point without having to find the function. In this section we prove a theorem which meets the first requirement—and, in the course of the proof, we succeed in meeting the second, too, so we incorporate the value of $\frac{dy}{dx}$ into the statement of the theorem.

Theorem 3.1. *Let* $f(x, y)$ *be a continuous function of* x *and* y. *We suppose that* $f(a, b)=0$, *that* f *is differentiable at* (a, b) *and that* $\frac{\partial f}{\partial y}$ *is continuous and non-zero at* (a, b).

23

Under these assumptions there is a unique continuous function $y=g(x)$, *defined for all* x *sufficiently near to* a, *which satisfies the equation* $f(x, y)=0$ *identically and such that* $g(a)=b$. *Moreover* $g(x)$ *is differentiable at* $x=a$, *and*

$$g'(a)= -\frac{f_x(a, b)}{f_y(a, b)}.$$

Before proving this theorem we try to clarify its meaning by an example and a few remarks. Consider $f(x, y)=x^2+y^2-1=0$. Then f satisfies the conditions of the theorem at $(0, 1)$ and $g(x)=\sqrt{(1-x^2)}$. Similarly f satisfies the conditions at $(0, -1)$ but now $g(x)= -\sqrt{(1-x^2)}$. On the other hand we cannot find a *unique* function g such that $g(1)=0$, since either of the functions mentioned above would serve. This is not surprising since $f_y(1, 0)=0$.

To say that $y=g(x)$ satisfies $f(x, y)=0$ identically is just to say that $f(x, g(x))$ is identically zero; thus in our example $x^2+(1-x^2)-1$ is identically zero. To say that there is, near $x=a$, a *unique* continuous function $y=g(x)$ satisfying $f(x, y)=0$ with $g(a)=b$ is to say that if $g_1(x), g_2(x)$ are two such functions, then $g_1(x)=g_2(x)$ for *all* x sufficiently near to a. Notice that $g(x)$ may only be defined near to $x=a$; in our example above $g(x)$ is not defined for $|x|>1$.

We now leave the example and make two remarks about the role played in the proof by the hypothesis of continuity of the functions f and f_y. We will invoke, in fact, the following two fundamental properties of continuous functions. The first may be described as the property of '*gaplessness*'; thus if $F(x)$ is a continuous function of x and $F(a)=p$, $F(b)=q$, and if r is any number between p and q then, for *some* c between a and b, $F(c)=r$. The second feature of a continuous function which we invoke is the property of the *persistence of inequalities*. Thus suppose that $u(x, y)$ is a continuous function and that $u(a, b)>c$, then the inequality $u(x, y)>c$ holds for all (x, y) sufficiently near to (a, b). For let $u(a, b)-c=d>0$. Then, since u is continuous,

$|u(x, y) - u(a, b)| < \dfrac{d}{2}$ if (x, y) is close enough to (a, b), and,

for such (x, y), $u(x, y) - c > \dfrac{d}{2}$ so that, certainly, $u(x, y) > c$.

We now prove the theorem. Since $f_y(a, b) \neq 0$ we may, without real loss of generality, suppose that $f_y(a, b) > 0$ (otherwise replace f by $-f$). It follows from the property of persistence of inequalities that $f_y(x, y) > 0$ for all (x, y) close enough to (a, b) and *we confine attention henceforth to such a neighbourhood of* (a, b). Since $f_y(a, b) > 0$ it follows that $f(a, b+k) > 0$ and $f(a, b-k) < 0$ for k sufficiently small and positive, say $0 < k \leqslant \delta$. We again invoke the persistence of inequalities to infer that for each such k there is a positive number $\rho(k)$, depending on k, such that $f(a+h, b+k) > 0$ and $f(a+h, b-k) < 0$ if $|h| \leqslant \rho(k)$. We emphasize, by writing $\rho(k)$, that $\rho(k)$ depends on k; let us write ρ for $\rho(\delta)$.

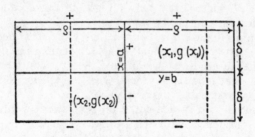

FIG. 2

To summarize our progress, we know that in Fig. 2 $f_y(x, y) > 0$ throughout the big rectangle, $f(x, y) > 0$ along the top horizontal and $f(x, y) < 0$ along the bottom horizontal. We also know (though we will only use this information later) that $f(a, b+y) > 0$ for $0 < y \leqslant \delta$, $f(a, b-y) < 0$ for $0 < y \leqslant \delta$.

Now choose x in $|x-a| \leqslant \rho$. Since $f(x, b+\delta) > 0$ and $f(x, b-\delta) < 0$, it follows from the property of gaplessness that there is a number y in $|y-b|<\delta$ for which $f(x, y)=0$. Moreover, there is only one such y; for if $f(x, y')=0$ and $f(x, y'')=0$ with $|y'-b|<\delta$, $|y''-b|<\delta$, then, for some y between y' and y'', $f_y(x, y)=0$ (Rolle's Theorem, DC, Theorem 3.2), contrary to the fact that $f_y(x, y)>0$ throughout the rectangle. Thus for each x in $|x-a|\leqslant\rho$ there is a unique y in $|y-b|<\delta$ satisfying the equation $f(x, y)=0$. We call this solution $g(x)$, so that $y=g(x)$ is a well-defined function of x in $|x-a|\leqslant\rho$ and plainly $g(a)=b$.

We next prove that the function $g(x)$ is continuous in $|x-a|<\rho$. Choose any a' in $|x-a|<\rho$ and let $g(a')=b'$. We prove $g(x)$ continuous at $x=a'$. This means that we must show that $g(x)$ remains as near as we like to b' if x remains sufficiently near to a'. Now $b+\delta>b'>b-\delta$. Thus if δ' is small we may suppose that the strip $b'+\delta'\geqslant y\geqslant b'-\delta'$ lies inside the strip $b+\delta\geqslant y\geqslant b-\delta$. Then the argument whereby we obtained $g(x)$ may be repeated† to produce an interval $|x-a'|\leqslant\rho'$ such that $|g(x)-b'|<\delta'$ if $|x-a'|\leqslant\rho'$. This establishes the continuity of g.

Summing up, we have shown that $y=g(x)$ is the unique function satisfying $f(x, y)=0$, $g(a)=b$, defined in $|x-a|\leqslant\rho$ and taking values in $|y-b|<\delta$; and that it is continuous. We conclude that it is the *unique continuous function* defined in $|x-a|\leqslant\rho$ and satisfying $f(x, y)=0$, $g(a)=b$. For if $g_1(x)$ were another, really distinct from $g(x)$, there would be‡ a biggest number x_0 with the property that $|x_0-a|<\rho$ and $g(x)=g_1(x)$ for $a\leqslant x\leqslant x_0$, or a smallest number x_0 with $|x_0-a|<\rho$ and $g(x)=g_1(x)$ for $a\geqslant x\geqslant x_0$. At x_0, $g_1(x)$ would have to 'jump' from $g_1(x_0)$, in the interior of the rectangle, over the boundary and this is impossible.

It remains to prove that $g(x)$ is differentiable at $x=a$ and to find the differential coefficient. Since $f(x, y)$ is differentiable at (a, b) and $f(a, b)=0$, (2.3) yields

† Notice that $f(a', y)>0$ if $b'<y\leqslant b+\delta$, $f(a', y)<0$ if $b'>y\geqslant b-\delta$.
‡ This fact about continuous functions, while intuitively obvious once its meaning is understood, is, like the property of 'gaplessness', by no means easy to prove. We are content to state such facts; proofs may be found, e.g. in Phillips's *Analysis*, Ch. 3.

$$f(a+h, b+k)=h(f_x(a, b)+\epsilon)+k(f_y(a, b)+\eta),$$

where ϵ, $\eta \to 0$ as h, $k \to 0$. Take $0 < |h| < \rho$ and put $k=g(a+h)-g(a)$. Then $f(a+h, b+k)=0$, so that

$$0=f_x(a, b)+\epsilon+\frac{g(a+h)-g(a)}{h}(f_y(a, b)+\eta); \quad (3.2)$$

notice that in (3.2) the symbols ϵ, η stand for the result of substituting for k in $\epsilon(h, k)$, $\eta(h, k)$, and are functions of h only. Since g is continuous, $g(a+h)-g(a)\to 0$ as $h\to 0$ so that in (3.2), ϵ, $\eta \to 0$ as $h\to 0$. Thus, since $f_y(a, b)>0$, it follows that $f_y(a, b)+\eta>0$ if h is sufficiently small, so that we may write

$$\frac{g(a+h)-g(a)}{h}=-\frac{f_x(a, b)+\epsilon}{f_y(a, b)+\eta},$$

whence

$$\lim_{h\to 0}\frac{g(a+h)-g(a)}{h}=-\frac{f_x(a, b)}{f_y(a, b)}.$$

This shows that $g(x)$ is differentiable at $x=a$ and establishes the fundamental formula

$$g'(a)=-\frac{f_x(a, b)}{f_y(a, b)}. \quad (3.3)$$

This completes the proof of the theorem. For most readers the principal conclusion of the theorem is to be found in the formula (3.3). This formula should remain in the reader's mind when all recollection of the proof of the theorem has faded. To consolidate we immediately give an example.

Example 3.1. Let (a, b) lie on the circle $x^2+y^2-c^2=0$. Then, putting $f(x, y)=x^2+y^2-c^2$ we have $f_x(a, b)=2a$, $f_y(a, b)=2b$ so that, by (3.3),

$$\frac{dy}{dx}=-\frac{a}{b},$$

at (a, b). Thus $\frac{dy}{dx}=-\frac{x}{y}$ for the general point (x, y) on the circle. Of course we must exclude the points $(\pm c, 0)$.

27

Theorem 3.1 generalizes in evident fashion to functions of more than two variables. In what follows we shall feel free to apply it in this more general form, especially to functions of three variables. More explicitly, if $f(x, y, z)$ is a well-behaved function and $f(a, b, c)=0$, $f_z(a, b, c)\neq0$, there is a unique function $z=g(x, y)$ satisfying $f(x, y, z)=0$, $g(a, b)=c$, and $g_x=-f_x/f_z$, $g_y=-f_y/f_z$.

2. DERIVATIVES INVOLVING IMPLICIT FUNCTIONS

The reader should notice that, granted Theorem 3.1, formula (3.3) is readily deducible from (2.7). For, as pointed out, to say that $y=g(x)$ satisfies $f(x, y)=0$ identically is to say that if $z=f(x, y)$ and if $y=g(x)$ then z, as a function of x only, is identically zero. Thus, by (2.7),

$$0=\frac{\partial f}{\partial x}+\frac{\partial f}{\partial y}\frac{dy}{dx},$$

whence (3.3) follows.

As an example we use the same technique to compute $\frac{d^2y}{dx^2}$.

Example 3.2. If x, y are related by the equation $f(x, y)=0$, then

$$\frac{d^2y}{dx^2}=-\frac{f_x^2 f_{yy}-2f_x f_y f_{xy}+f_y^2 f_{xx}}{f_y^3}.$$

We have $\frac{dy}{dx}=-\frac{f_x}{f_y}$; thus to compute $\frac{d^2y}{dx^2}$, we must calculate $\frac{d}{dx}f_x$ and $\frac{d}{dx}f_y$. Expressing (2.7) operationally (see example 2.5), we have $\frac{d}{dx}=\frac{\partial}{\partial x}+\frac{dy}{dx}\frac{\partial}{\partial y}$ (the argument, or object of the operation, is any function of x and y, y being a given function of x). Applying this to the function f_x, we have $\frac{df_x}{dx}=f_{xx}+\frac{dy}{dx}f_{xy}=f_{xx}-\frac{f_x}{f_y}f_{xy}$, and similarly applying it to the function† f_y, $\dfrac{df_y}{dx} = f_{xy} + \dfrac{dy}{dx} f_{yy} = f_{xy} - \dfrac{f_x}{f_y} f_{yy}$.

Thus
$$\frac{d^2y}{dx^2} = -\frac{f_y(df_x/dx) - f_x(df_y/dx)}{f_y^2}$$
$$= -\frac{f_y(f_y f_{xx} - f_x f_{xy}) - f_x(f_y f_{xy} - f_x f_{yy})}{f_y^3},$$

and the result follows. We have, of course, assumed in this calculation that $f_y \neq 0$ at the point at which $\dfrac{d^2y}{dx^2}$ is being evaluated.

We continue this section by discussing a type of situation of frequent occurrence in thermodynamical problems. Suppose three variables‡ x, y, z are connected by an equation $g(x, y, z) = 0$ and that a fourth variable w is given as a function of x, y, z say,

$$w = f(x, y, z). \tag{3.4}$$

By eliminating z, we may express w as a function of x and y, and hence form $\left(\dfrac{\partial w}{\partial x}\right)_y$; similarly, we may form $\left(\dfrac{\partial w}{\partial x}\right)_z$. These two are of course different from each other and different from $\left(\dfrac{\partial w}{\partial x}\right)_{y,\,z}$ calculated directly from (3.4). It follows from (2.7) that, in fact,

$$\left(\frac{\partial w}{\partial x}\right)_y = \left(\frac{\partial w}{\partial x}\right)_{y,\,z} + \left(\frac{\partial w}{\partial z}\right)_{x,\,y} \left(\frac{\partial z}{\partial x}\right)_y. \tag{3.5}$$

Notice that, effectively, this is just a 'two-variable' formula since y is held fixed in all terms.

Example 3.3. If $U = f(T, p, v)$, where $g(T, p, v) = 0$, prove that

$$\left(\frac{\partial U}{\partial p}\right)_T = \left(\frac{\partial U}{\partial p} \cdot \frac{\partial T}{\partial v} - \frac{\partial U}{\partial v} \cdot \frac{\partial T}{\partial p}\right) \Big/ \frac{\partial T}{\partial v}.$$

† Recall that $f_{xy} = f_{yx}$; we will always write f_{xy} for either of these.
‡ For example, temperature, pressure, and volume.

On the right-hand side we have reverted to the abbreviated notation: $\dfrac{\partial U}{\partial p}$, $\dfrac{\partial U}{\partial v}$ simply refer to the function f and $\dfrac{\partial T}{\partial p}$, $\dfrac{\partial T}{\partial v}$ refer to the expression of T as a function of p, v, obtainable from $g(T, p, v) = 0$.

From (3.5) we immediately infer that

$$\left(\frac{\partial U}{\partial p}\right)_T = \frac{\partial U}{\partial p} + \frac{\partial U}{\partial v}\frac{\partial v}{\partial p}, \text{ where } \frac{\partial v}{\partial p} \text{ means } \left(\frac{\partial v}{\partial p}\right)_T.$$

Now $\left(\dfrac{\partial v}{\partial p}\right)_T$ is computed from $g(T, p, v) = 0$, T being held fixed. Thus Theorem 3.1 may be applied, in the following way. Let $T = h(p, v)$ be the solution of $g(T, p, v) = 0$. Then $\dfrac{\partial T}{\partial p} = h_p$, $\dfrac{\partial T}{\partial v} = h_v$. On the other hand, holding T fixed, we have

$$\left(\frac{\partial v}{\partial p}\right)_T = -\frac{h_p}{h_v}, \tag{3.6}$$

by Theorem 3.1 (notice that we may replace '$f(x, y) = 0$' by '$f(x, y) = c$' in the statement of that theorem). Thus

$$\frac{\partial v}{\partial p} = \left(\frac{\partial v}{\partial p}\right)_T = -\frac{\partial T}{\partial p} \bigg/ \frac{\partial T}{\partial v},$$

and the required formula for $\left(\dfrac{\partial U}{\partial p}\right)_T$ is obtained. The reader is advised to check the steps of the calculation in a particular case, say, for $U = e^{-\alpha T}p^{\beta}v^{\gamma}$, where $pv = kT$. He may also find it valuable to solve this example using the extension of Theorem 3.1 to a function of three variables.

3. JACOBIANS, INVERSE FUNCTIONS, AND FUNCTIONAL DEPENDENCE

A particular case of Theorem 3.1 is provided by the equation $x = h(y)$, where $h(b) = a$. By putting $f(x, y) = h(y) - x$ we deduce:

Corollary 3.1. *If* h(y) *is differentiable near* y=b *and* h′(y) *is continuous with* h′(b) ≠0, *then, for* x *near* a, *there is a unique continuous function* y=g(x) *with* b=g(a) *such that* h(g(x))=x. *Moreover,* g(x) *is differentiable at* x=a *and* $g'(a) = \dfrac{1}{h'(b)}$.

The function $g(x)$ is called the *inverse* of the function $h(y)$ (see DC, p. 19).

We now consider the generalization of this result to functions of two variables. We suppose given two differentiable functions $u=f(x, y)$, $v=g(x, y)$ such that $u_0=f(x_0, y_0)$, $v_0=g(x_0, y_0)$, and we ask whether there exist unique differentiable functions $x=F(u, v)$, $y=G(u, v)$ satisfying the equations $u=f(x, y)$, $v=g(x, y)$ identically and such that $x_0=F(u_0, v_0)$, $y_0=G(u_0, v_0)$. To answer this question we introduce a definition.

Definition 3.1. *The* Jacobian *of the transformation* u=f(x, y), v=g(x, y) *is the determinant*†

$$\frac{\partial(u, v)}{\partial(x, y)} = \begin{vmatrix} \dfrac{\partial u}{\partial x} & \dfrac{\partial u}{\partial y} \\[2mm] \dfrac{\partial v}{\partial x} & \dfrac{\partial v}{\partial y} \end{vmatrix}.$$

The matrix $\begin{bmatrix} \dfrac{\partial u}{\partial x} & \dfrac{\partial u}{\partial y} \\[2mm] \dfrac{\partial v}{\partial x} & \dfrac{\partial v}{\partial y} \end{bmatrix}$ is sometimes referred to as the *Jacobian matrix* of the transformation.

We may now state the theorem.

Theorem 3.2. *The functions* x=F(u, v), y=G(u, v) *exist if* $\dfrac{\partial u}{\partial x}, \dfrac{\partial u}{\partial y}, \dfrac{\partial v}{\partial x}, \dfrac{\partial v}{\partial y}$ *are continuous at* (x_0, y_0) *and* $\dfrac{\partial(u, v)}{\partial(x, y)} \neq 0$ *at* (x_0, y_0).

† See Cohn, *Linear Equations*, Ch. 5.

We sketch the proof. Since $\dfrac{\partial(u, v)}{\partial(x, y)} \neq 0$ at (x_0, y_0) either $\dfrac{\partial u}{\partial x}$ or $\dfrac{\partial u}{\partial y}$ is non-zero at (x_0, y_0). Suppose that $\dfrac{\partial u}{\partial x} \neq 0$; we may then apply Theorem 3.1 to obtain x as a function of u and y, say $x = h(u, y)$ and (see (3.6))

$$\frac{\partial h}{\partial y} = \left(\frac{\partial x}{\partial y}\right)_u = -\frac{\partial f / \partial y}{\partial f / \partial x}.$$

Substituting for x in the equation $v = g(x, y)$ we get $v = g(h(u, y), y)$ and, again applying Theorem 3.1, we deduce that we may solve for y as a function of u and v provided $\left(\dfrac{\partial v}{\partial y}\right)_u \neq 0$; but, by Theorem 2.2,

$$\begin{aligned}
\left(\frac{\partial v}{\partial y}\right)_u &= \frac{\partial g}{\partial x}\left(\frac{\partial x}{\partial y}\right)_u + \frac{\partial g}{\partial y} \\
&= \left(\frac{\partial f}{\partial x}\frac{\partial g}{\partial y} - \frac{\partial f}{\partial y}\frac{\partial g}{\partial x}\right) \Big/ \frac{\partial f}{\partial x} \\
&= \frac{\partial(u, v)}{\partial(x, y)} \Big/ \frac{\partial u}{\partial x} \neq 0.
\end{aligned}$$

Thus $y = G(u, v)$; substituting back in $x = h(u, y)$ we get $x = F(u, v)$. The uniqueness and differentiability of F and G now follow substantially as in Theorem 3.1.

Example 3.4. Let $u = \dfrac{x^2}{y}$, $v = \dfrac{y^2}{x}$. Then

$$\frac{\partial(u, v)}{\partial(x, y)} = \begin{vmatrix} 2x/y & -x^2/y^2 \\ -y^2/x^2 & 2y/x \end{vmatrix} = 3.$$

Thus we may solve for x and y at any point off the coordinate axes. (Actually we find $x = u^{2/3}v^{1/3}$, $y = u^{1/3}v^{2/3}$ and these functions are differentiable away from $u = 0$, $v = 0$.)

Theorem 3.2 naturally suggests the question: what conclusion can we draw if $\dfrac{\partial(u, v)}{\partial(x, y)} = 0$ near (x_0, y_0)? We remark that, for functions of one variable, the vanishing of $\dfrac{du}{dx}$ near x_0 simply means that u is constant near x_0. The generalization is:

Theorem 3.3. *If* $\dfrac{\partial(u, v)}{\partial(x, y)} = 0$ *at all points near* (x_0, y_0), *and if* u_x, v_x, u_y, v_y *are continuous at* (x_0, y_0) *and not all zero there,*† *then the functions* u, v *satisfy a functional relation* $\phi(u, v) = 0$ *near* (x_0, y_0).

Before proving the theorem we exemplify it. Let $u = x^2 y^2$, $v = \log x + \log y$. Then if $x_0 \neq 0$, $y_0 \neq 0$ we may evaluate $\dfrac{\partial(u, v)}{\partial(x, y)}$ at (x_0, y_0) to get

$$\begin{vmatrix} 2x_0 y_0^2 & 2x_0^2 y_0 \\ 1/x_0 & 1/y_0 \end{vmatrix} = 0;$$

thus $\dfrac{\partial(u, v)}{\partial(x, y)} = 0$ near any point off the co-ordinate axes. In fact u, v satisfy the relation $\log u - 2v = 0$.

We now prove the theorem. Without real loss of generality we may suppose that $\dfrac{\partial u}{\partial x} \neq 0$ at (x_0, y_0); we may then argue essentially as in the proof of Theorem 3.2. That is, we first show that $x = h(u, y)$ so that $v = g(h(u, y), y)$ and then prove

$$\left(\frac{\partial v}{\partial y} \right)_u = \frac{\partial(u, v)}{\partial(x, y)} \bigg/ \frac{\partial u}{\partial x}.$$

But now $\dfrac{\partial(u, v)}{\partial(x, y)} = 0$ so that $\left(\dfrac{\partial v}{\partial y} \right)_u = 0$ and v is a function of u only. Thus we have a functional relation $v - \psi(u) = 0$ and the theorem is proved.

Theorem 3.3 has an evident converse: *if the functions* u, v *satisfy* $\phi(u, v) = 0$ *and if* ϕ_u, ϕ_v *are continuous at* (u_0, v_0) *and*

---

† We include this rather strong condition to avoid an awkward discussion of stationary points (see Ch. 4).

not both zero there, then $\dfrac{\partial(u, v)}{\partial(x, y)}$ *vanishes near* (x_0, y_0). We leave the (fairly easy) proof to the reader.

Jacobians play a fundamental role not only in the theory of implicit functions, as demonstrated, but also in other branches of the calculus, and particularly in the integration of functions of several variables.[†] We therefore close by drawing attention to certain important properties they possess.

Theorem 3.4. *Suppose* u, v *are differentiable functions of* x, y *and* x, y *are differentiable functions of* ξ, η. *Then*

$$\frac{\partial(u, v)}{\partial(\xi, \eta)} = \frac{\partial(u, v)}{\partial(x, y)} \frac{\partial(x, y)}{\partial(\xi, \eta)}.$$

The proof is a pleasant exercise in applying Theorem 2.2. We first recall that if

$$A = \begin{pmatrix} a_{11} & a_{12} \\ a_{21} & a_{22} \end{pmatrix}, \qquad B = \begin{pmatrix} b_{11} & b_{12} \\ b_{21} & b_{22} \end{pmatrix}$$

are matrices with determinants $|A|$, $|B|$, then

$$AB = \begin{pmatrix} a_{11}b_{11} + a_{12}b_{21} & a_{11}b_{12} + a_{12}b_{22} \\ a_{21}b_{11} + a_{22}b_{21} & a_{21}b_{12} + a_{22}b_{22} \end{pmatrix}$$

and[‡]

$$|AB| = |A|\ |B|. \tag{3.7}$$

Thus if $A = \begin{pmatrix} u_x & u_y \\ v_x & v_y \end{pmatrix}$, $B = \begin{pmatrix} x_\xi & x_\eta \\ y_\xi & y_\eta \end{pmatrix}$ then, by Theorem 2.2,

$$AB = \begin{pmatrix} u_x x_\xi + u_y y_\xi & u_x x_\eta + u_y y_\eta \\ v_x x_\xi + v_y y_\xi & v_x x_\eta + v_y y_\eta \end{pmatrix} = \begin{pmatrix} u_\xi & u_\eta \\ v_\xi & v_\eta \end{pmatrix}$$

and the theorem follows by applying (3.7).

Corollary 3.2. *If* u, v *are differentiable functions of* x, y *with non-vanishing Jacobian, then*

---

† See, for example, E. C. Phillips, *Analysis*, C.U.P., Ch. XI, sec. 8.
‡ See Cohn, *Linear Equations*, Ch. 5.

$$\frac{\partial(u, v)}{\partial(x, y)} \cdot \frac{\partial(x, y)}{\partial(u, v)} = 1.$$

For then x, y are differentiable functions of u, v and we apply Theorem 3.4 with $u = \xi$, $v = \eta$.

This corollary is very useful for calculating $\frac{\partial(x, y)}{\partial(u, v)}$ without having to compute any of the partial derivatives x_u, x_v, y_u, y_v. (It is, of course, the generalization to two variables of the formula $\frac{dy}{dx} \cdot \frac{dx}{dy} = 1$.)

Example 3.5. If $u = \frac{x^2}{y}$, $v = \frac{y^2}{x}$ then $\frac{\partial(x, y)}{\partial(u, v)} = \frac{1}{3}$. For we have seen (example 3.4) that $\frac{\partial(u, v)}{\partial(x, y)} = 3$. The reader should check this result by direct verification.

Exercises

1. Find $\frac{dy}{dx}$ when x, y are connected by the equation
   (i) $xy + x + y = 1$;
   (ii) $x \cos y - y \cos x = c$.
Find the gradient to the curve $e^{x \sin y} - 2e^{x \cos y} = e^x$ at the point $(\log \frac{1}{3}, 0)$.

2. If $y^3 + 3x^2 y = c$, prove that $\frac{d^2y}{dx^2} = \frac{2c(x^2 - y^2)}{(x^2 + y^2)^3}$.

3. If $f(x, y, z) = 0$, prove that $\left(\frac{\partial x}{\partial y}\right)_z \left(\frac{\partial y}{\partial z}\right)_x \left(\frac{\partial z}{\partial x}\right)_y = -1$.

Verify this if $\frac{x}{y} + \frac{y}{z} + \frac{z}{x} = c$.

4. If $x + y + z + w = 0$, $x^2 + y^2 + z^2 + w^2 = 2$, $x^3 + y^3 + z^3 + w^3 = 3$, prove that
$$\frac{dy}{dx} = -\frac{1 - 3x^2 - 2xy - y^2}{1 - x^2 - 2xy - 3y^2}.$$

5. If $u(x, y, z)=0$, $v(x, y, z)=0$, prove that
$$\frac{dy}{dx}=-\frac{u_z v_x - u_x v_z}{u_z v_y - u_y v_z}.$$
What does this formula become if there is a function $w=w(x, z)$ such that u and v are each functions of y and w? Interpret your result and demonstrate it in the case $u=xy+yz+y-3$, $v=x-y+z-1$.

6. If $x=r\cos\theta$, $y=r\sin\theta$, find $\dfrac{\partial(x, y)}{\partial(r, \theta)}$. Calculate $\dfrac{\partial(r, \theta)}{\partial(x, y)}$ and verify that $\dfrac{\partial(x, y)}{\partial(r, \theta)} \cdot \dfrac{\partial(r, \theta)}{\partial(x, y)}=1$.

7. If $u=e^{x+y}\cos(x-y)$, $v=e^{x+y}\sin(x-y)$, find $\dfrac{\partial(x, y)}{\partial(u, v)}$.

8. Extend Definition 3.1 and the statements of Theorems 3.1 to 3.4 to functions of three variables.

9. (i) Find $\dfrac{\partial(x, y, z)}{\partial(r, \theta, \phi)}$ if $x=r\sin\theta\cos\phi$, $y=r\sin\theta\sin\phi$, $z=r\cos\theta$.

(ii) Find $\dfrac{\partial(x, y, z)}{\partial(r, \theta, z)}$ if $x=r\cos\theta$, $y=r\sin\theta$.

(iii) Find $\dfrac{\partial(x, y, z)}{\partial(u, v, w)}$ if $x=\dfrac{u^2}{v}$, $y=\dfrac{v^2}{w}$, $z=\dfrac{w^2}{u}$.

10. Show in each of the following cases that the functions f, g, h satisfy a functional relation:

(i) $f=x+y+z$, $g=x^2+y^2+z^2+2yz$,
$\quad h=x^3+y^3+z^3+3yz(y+z)$;

(ii) $f=x+2z$, $g=\dfrac{x-y}{y+2z}$, $h=x^2-4z^2-2xy-4yz$;

(iii) $f=e^{2x+z}(x-y)$; $g=e^{2y+z}(x+y+z)$,
$\quad h=z^2+4xy+2xz+2yz$.

11. Find the normal to the ellipsoid $\dfrac{x^2}{a^2}+\dfrac{y^2}{b^2}+\dfrac{z^2}{c^2}=1$ at the point (x_0, y_0, z_0). Prove that in general six normals may be drawn to the ellipsoid from a point inside it.

CHAPTER FOUR
Maxima and Minima

---

1. MEAN VALUE THEOREM AND TAYLOR'S THEOREM

The theory of maxima and minima for functions of several variables follows closely the lines of the theory for functions of one variable described in DC. Thus there are generalizations of the Mean Value Theorem (here abbreviated, as in DC, to MVT) and Taylor's Theorem, and criteria for the existence of maxima or minima may be deduced from them. The deductions are, however, somewhat more sophisticated than in the simpler case of a single variable. As in previous chapters we state the main results only for functions of two variables, but the reader should be able to supply the extension to several variables.

Theorem 4.1 (MVT). *If* f(x, y) *is differentiable, then there exists a number* θ, *satisfying* $0 < \theta \leqslant 1$ *and depending on* a, b, h, k *such that*

$$f(a+h, b+k) = f(a, b) + hf_x(a+\theta h, b+\theta k) + kf_y(a+\theta h, b+\theta k).$$

We prove this by the trick of reducing the problem to one involving only a function of one variable.

Let $F(t) = f(a+ht, b+kt)$. Then $F(t)$ is differentiable and, by (2.6),

$$F'(t) = hf_x(a+ht, b+kt) + kf_y(a+ht, b+kt).$$

Now, by MVT for one variable, $F(1) = F(0) + F'(\theta)$, for some θ satisfying $0 < \theta < 1$. This is exactly the assertion of the theorem.

In a similar way, we may deduce Taylor's Theorem for $f(x, y)$ by writing down the Taylor expansion of $F(t)$ and interpreting the successive derivatives of $F(t)$ in terms of the partial derivatives of $f(x, y)$. We content ourselves with obtaining the second-order terms and state the result as

Theorem 4.2. Let $f(x, y)$ *be a function with continuous partial derivatives of the first and second order. Then*

$f(a+h, b+k)=f(a, b)+hf_x(a, b)+kf_y(a, b)$
$+\frac{1}{2}\{h^2 f_{xx}(a+\theta h, b+\theta k)+2hkf_{xy}(a+\theta h, b+\theta k)$
$+k^2 f_{yy}(a+\theta h, b+\theta k)\}$,

for some θ, $0<\theta<1$.

To prove this we must evaluate $F''(t)$, where $F(t)=f(a+ht, b+kt)$; indeed, in view of Taylor's Theorem for one variable, we are asserting precisely that
$F''(t)=h^2 f_{xx}(a+ht, b+kt)+2hkf_{xy}(a+ht, b+kt)$
$+k^2 f_{yy}(a+ht, b+kt)$.

Now $\dfrac{d}{dt}=h\dfrac{\partial}{\partial x}+k\dfrac{\partial}{\partial y}$. Thus $F''(t)=\dfrac{d}{dt}F'(t)=\dfrac{d}{dt}(hf_x+kf_y)$

$=\left(h\dfrac{\partial}{\partial x}+k\dfrac{\partial}{\partial y}\right)(hf_x+kf_y)=h^2 f_{xx}+2hkf_{xy}+k^2 f_{yy}$, and the theorem is proved.

The reader should compare Theorem 4.2 with (2.5); there the difference $f(a+h, b+k)-\{f(a, b)+hf_x(a, b)+kf_y(a, b)\}$ was expressed as $\epsilon\sqrt{(h^2+k^2)}$ where $\epsilon\to0$ as $h, k\to0$. Under the stronger hypotheses of the theorem we are able to give a more explicit expression for ϵ.

If we replace $a+h$ by x and $b+h$ by y, we may interpret Theorem 4.2 as showing how closely we may approximate to $f(x, y)$ near (a, b) by the linear function
$$f(a, b)+(x-a)f_x(a, b)+(y-b)f_y(a, b).$$

2. MAXIMA AND MINIMA

We now investigate maxima and minima for functions of two (or more) variables. As in DC we are concerned with local maxima and minima and as there we issue the warning that our methods will not pick out maxima or minima on the boundary of a region under consideration. Thus to say that $f(x, y)$ has a maximum at (a, b) is to say that $f(a, b) >$ $f(x, y)$ for all (x, y) sufficiently near to (a, b). We suppose f and its derivatives (at least of the second order) to be continuous.

If $f(x, y)$ has a maximum (or minimum) at (a, b), then certainly $f(x, b)$ has a maximum (or minimum) at $x=a$. Thus, by the theory for functions of one variable, $f_x(a, b)=0$; similarly $f_y(a, b)=0$. *These then are necessary conditions*. However, it may well happen that (a, b) is not a maximum (or minimum) of $f(x, y)$ although it is so if we approach (a, b) along $x=a$ or $y=b$. For example, consider the function

$$f(x, y)=2(x+y)^2 - x^2 - y^2.$$

On the x-axis, $y=0$ and $f(x, 0)=x^2$. Thus $(0, 0)$ is a minimum on the x-axis. Similarly $f(0, y)=y^2$, so that $(0, 0)$ is a minimum on the y-axis. On the other hand, if we approach the origin along $x+y=0$, we have

$$f(x, -x)=-2x^2,$$

so that $(0, 0)$ is actually a maximum along this line.

We call a point (a, b) at which f_x and f_y vanish a *stationary* point and use Theorem 4.2 to obtain sufficient conditions for maxima or minima at stationary points.

We first need a lemma and its converse from algebra. A *quadratic form* in the variables $x_1, \ldots, x_n$ is a polynomial in these variables all of whose terms are of the second degree; thus

$a_1 x_1^2 + \ldots + a_n x_n^2 + 2a_{12} x_1 x_2 + 2a_{13} x_1 x_3 + \ldots + 2a_{n-1,n} x_{n-1} x_n,$

where $a_1, \ldots, a_n, a_{12}, \ldots, a_{n-1,n}$ are real constants, is the general quadratic form in $x_1, \ldots, x_n$. Here we consider only quadratic forms in two variables x, y and prove†

Lemma 4.1. *If* $q^2 < pr$, *the quadratic form* $px^2 + 2qxy + ry^2$ *vanishes only at the origin and, elsewhere, has the same sign as* p.

Suppose $q^2 < pr$. Then certainly $p \neq 0$ and $px^2 + 2qxy + ry^2$

† This lemma and Lemma 4·2 do generalize to forms in more than two variables, but the generalizations are not obvious.

$=\frac{1}{p}\{(px+qy)^2+(pr-q^2)y^2\}$. Since the expression in the curly brackets is positive unless $px+qy=0$ and $y=0$, i.e., unless $(x, y)=(0, 0)$, it follows that $px^2+2qxy+ry^2$ has the same sign as p except at $(0, 0)$. This proves the lemma.

We now examine the case $q^2 \geqslant pr$; suppose first that $q^2=pr$. Then if $p \neq 0$, $px^2+2qxy+ry^2=\frac{1}{p}(px+qy)^2$ and can therefore be zero even if $(x, y) \neq (0, 0)$. In fact, this happens all along the line $px+qy=0$ and therefore arbitrarily close to $(0, 0)$. The same conclusion follows even if $p=0$; the reader should supply the proof.† Now suppose $q^2>pr$. If $p \neq 0$, we see as before that $px^2+2qxy+ry^2=\frac{1}{p}\{(px+qy)^2-(q^2-pr)y^2\}$. Thus there are points, arbitrarily close to $(0, 0)$, at which the form takes opposite signs; in fact on $y=0$, it has the same sign as p, except at $(0, 0)$, and on $px+qy=0$ it has the opposite sign to p, except at $(0, 0)$. Even if $p=0$, we can still prove (and the reader should) that there are two lines through $(0, 0)$ along which the form takes opposite signs. Summing up our conclusions, we have:

Lemma 4.2. *If* $q^2=pr$, *the quadratic form* $px^2+2qxy+ry^2$ *takes the value* 0 *along a line through* $(0, 0)$. *If* $q^2>pr$, *the quadratic form takes opposite signs along two lines through‡* $(0, 0)$.

We now revert to the question of maxima and minima. We have, at a stationary point (a, b),

$f(a+h, b+k)-f(a, b)$
$=\frac{1}{2}\{h^2 f_{xx}(a+\theta h, b+\theta k)+2hk f_{xy}(a+\theta h, b+\theta k)$
$+k^2 f_{yy}(a+\theta h, b+\theta k)\}$,

where $0<\theta<1$.

---

† If $r \neq 0$, the form vanishes along $y=0$; if $r=0$ it vanishes everywhere.
‡ Except, of course, that it is zero at $(0, 0)$. Notice that Lemma 4.2 is a converse of Lemma 4.1 since it implies that if the form vanishes only at the origin and always has the same sign elsewhere then $q^2<pr$.

Let us use the notation $F(\xi, \eta; a', b')$ for the quadratic form (in the variables ξ, η)

$$\xi^2 f_{xx}(a', b') + 2\xi\eta f_{xy}(a', b') + \eta^2 f_{yy}(a', b').$$

Now suppose that $(f_{xy}(a, b))^2 < f_{xx}(a, b)f_{yy}(a, b)$ and that $f_{xx}(a, b) < 0$. Then by the assumed continuity of f_{xx}, f_{xy}, f_{yy} at (a, b) we infer from the property of the persistence of inequalities that $(f_{xy}(a+h', b+k'))^2 < f_{xx}(a+h', b+k') f_{yy}(a+h', b+k')$ and $f_{xx}(a+h', b+k') < 0$ for h', k' sufficiently small, say $|h'| < \delta$, $|k'| < \delta$. It follows that, for all (ξ, η) except $(0, 0)$, $F(\xi, \eta; a+h', b+k') < 0$ provided $|h'| < \delta$, $|k'| < \delta$; this is just an application of Lemma 4.1. Now, by (4.1), there is a number θ, $0 < \theta < 1$, such that

$$f(a+h, b+k) - f(a, b) = \tfrac{1}{2} F(h, k; a+\theta h, b+\theta k). \qquad (4.2)$$

We deduce immediately that, if $|h| < \delta$, $|k| < \delta$, then $f(a+h, b+k) < f(a, b)$ unless $h = k = 0$, so that $f(a, b)$ is a maximum. Arguing similarly if $(f_{xy}(a, b))^2 < f_{xx}(a, b) f_{yy}(a, b)$ and $f_{xx}(a, b) > 0$, we have proved:

Theorem 4.3. *The function* f(x, y) *has a maximum (minimum) at the stationary point* (a, b) *if*

$$(f_{xy}(a, b))^2 < f_{xx}(a, b)f_{yy}(a, b)$$

and f_{xx}(a, b) *is negative (positive).*

Next suppose that $(f_{xy}(a, b))^2 > f_{xx}(a, b)f_{yy}(a, b)$. Then by Lemma 4.2 there is a line $\alpha\xi + \beta\eta = 0$ at every point of which (except $(0, 0)$)

$$F(\xi, \eta; a, b) > 0$$

and another line $\alpha'\xi + \beta'\eta = 0$ at every point of which (except $(0, 0)$)

$$F(\xi, \eta; a, b) < 0.$$

It follows immediately from continuity considerations that there exists $\delta > 0$ such that $F(\xi, \eta; a', b') > 0$ if $\alpha\xi + \beta\eta = 0$ and $|a' - a| < \delta$, $|b' - b| < \delta$; and $F(\xi, \eta; a', b') < 0$ if $\alpha'\xi + \beta'\eta = 0$ and $|a' - a| < \delta$, $|b' - b| < \delta$. It follows now from (4.2) that if $|h| < \delta$, $|k| < \delta$, then

$$f(a+h, b+k) > f(a, b) \quad \text{if} \quad \alpha h + \beta k = 0,$$
$$f(a+h, b+k) < f(a, b) \quad \text{if} \quad \alpha' h + \beta' k = 0,$$

unless $h = k = 0$.

We have proved:

Theorem 4.4. *If* (a, b) *is a stationary point of the function* f(x, y) *and if*

$$(f_{xy}(a, b))^2 > f_{xx}(a, b)f_{yy}(a, b)$$

then f *has neither a maximum nor a minimum at* (a, b).

The situation described in this theorem may perhaps be best appreciated from a geometrical viewpoint. Consider, for example, the surface $z = x^2 - y^2$ (see Fig. 3). The origin is obviously a stationary point of the function z and $z_{xy} = 0$, $z_{xx} = 2$, $z_{yy} = -2$, so that the origin is neither a maximum nor a minimum. It is obviously a maximum looking along the plane $x = 0$ and a minimum looking along the plane $y = 0$. We call such a point a *saddle point*.

Reverting to our original problem, suppose finally that $(f_{xy}(a, b))^2 = f_{xx}(a, b)f_{yy}(a, b)$. Then there is a line through (a, b) along which the form is zero; along this line the sign of $f(a+h, b+k) - f(a, b)$ depends on the higher order terms in h, k of the Taylor expansion. As a rule it is better to examine this case by a special argument than to investigate the higher order terms of the Taylor expansion of $f(x, y)$.

Example 4.1. Find the maxima and minima of $x^4 + y^4 - 2x^2 + 4xy - 2y^2 - 1$. Put $f(x,y) = x^4 + y^4 - 2x^2 + 4xy - 2y^2 - 1$. Then $f_x = 4x^3 - 4x + 4y$, $f_y = 4y^3 + 4x - 4y$. Thus for stationary points we must solve $x^3 - x + y = 0$, $y^3 + x - y = 0$. Adding, $x^3 + y^3 = 0$, whence $x = -y$, so that $x^3 - 2x = 0$, $x = 0, \pm \sqrt{2}$, $y = 0, \mp \sqrt{2}$. Thus the stationary points are $(\sqrt{2}, -\sqrt{2})$, $(-\sqrt{2}, \sqrt{2})$, and (0, 0).

Now $f_{xx} = 12x^2 - 4$, $f_{xy} = 4$, $f_{yy} = 12y^2 - 4$. At $(\pm\sqrt{2}, \mp\sqrt{2})$, $f_{xx}f_{yy} - f_{xy}^2 = 384$. Thus $f(x, y)$ has minima at $(\pm\sqrt{2}, \mp\sqrt{2})$; its value at these points is -9. On the other

42

hand, at $(0, 0)$, $f_{xx}f_{yy}=f_{xy}^2$; since $f_{xx}<0$, $(0, 0)$ must be a maximum if it is either; but, on $x=y$, we have $f(x, x)=2x^4-1$, so that $(0, 0)$ is certainly not a maximum.

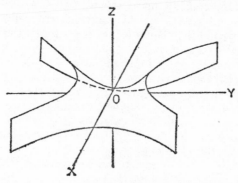

Fᴵɢ. 3

Theorem 4.3 may be generalized to functions of more than two variables. Since we do not propose here to generalize Lemma 4.1, we express the result as follows: We will say that the form $F= \sum\limits_{i,\,j=1}^{n} a_{ij}h_ih_j$ (F is a quadratic form in the variables $h_1, \ldots, h_n$ is *positive definite* if it is positive unless each h_i is zero; and we define *negative definite* similarly. Now let $f(x_1, \ldots, x_n)$ be a function of n variables $x_1, \ldots, x_n$ and let $a_{ij}=f_{x_ix_j}$, evaluated at $(c_1, \ldots, c_n)$. Then we may show exactly as in the case of functions of two variables that $f(x_1, \ldots, x_n)$ has a minimum (maximum) at $(c_1, \ldots, c_n)$ if $(c_1, \ldots, c_n)$ is a stationary point and F is positive (negative) definite. However, we will not use the criterion given above for determining whether a stationary point is a maximum or minimum in the case of functions of more than two variables. We remark, in passing, that $f(x, y, z)$ can only have a maximum at (a, b, c) if $f(x, y, c)$ has a maximum at (a, b, c) and that such an argument may well extablish the fact that (a, b, c) is not a maximum of $f(x, y, z)$.

3. MAXIMA AND MINIMA FOR FUNCTIONS WITH RE-STRAINTS

Hitherto in considering functions of two variables we have effectively supposed that the functions are defined in the plane (or part of the plane). It may happen that the functions are defined over a surface in space. Thus the problem may be to find maxima and minima of $f(x, y, z)$ when the variables x, y, z satisfy some relation $g(x, y, z)=0$. Now it may be convenient to eliminate one variable and thus regard the problem simply as one involving a function of two variables. Even where actual elimination is awkward, one may be able to introduce parameters (u, v) to describe the surface and thus reduce the problem to one involving a function of u and v. Thus, for example, to find maxima and minima of $yz+zx+xy$ on the sphere $x^2+y^2+z^2=1$, we may parametrize the sphere by $x=\sin \theta \cos \phi$, $y=\sin \theta \sin \phi$, $z=\cos \theta$. Then the problem is simply to find maxima and minima of $\sin \theta \cos \theta \sin \phi+\sin \theta \cos \theta \cos \phi+\sin^2\theta \sin \phi \cos \phi$, a function of the two variables θ, ϕ.

However it often happens that neither of the two procedures described above is convenient. We now describe a third procedure which is often the most convenient even where one or other of the given two is applicable. However, we will content ourselves with finding stationary points, laying down no preferred procedure for determining whether they are maxima or minima.

Thus the problem is to determine stationary points of $w=f(x, y, z)$ subject to $g(x, y, z)=0$. If we were to eliminate z, say, and then differentiate with respect to x, we would find (see (3.5))

$$\left(\frac{\partial w}{\partial x}\right)_y=\frac{\partial f}{\partial x}+\frac{\partial f}{\partial z}\frac{\partial z}{\partial x}$$

and, similarly,

$$\left(\frac{\partial w}{\partial y}\right)_x=\frac{\partial f}{\partial y}+\frac{\partial f}{\partial z}\frac{\partial z}{\partial y}.$$

Keeping y fixed in $g(x, y, z)=0$, we get from (3.3) that

$$\frac{\partial z}{\partial x}=-\frac{\partial g/\partial x}{\partial g/\partial z}; \text{ similarly, } \frac{\partial z}{\partial y}=-\frac{\partial g/\partial y}{\partial g/\partial z}.$$

Now at a stationary point $\left(\dfrac{\partial w}{\partial x}\right)_y=\left(\dfrac{\partial w}{\partial y}\right)_x=0$, whence

$$\frac{\partial f/\partial x}{\partial g/\partial x}=\frac{\partial f/\partial y}{\partial g/\partial y}=\frac{\partial f/\partial z}{\partial g/\partial z}. \tag{4.3}$$

Conversely, if (4.3) holds, then $\left(\dfrac{\partial w}{\partial x}\right)_y=\left(\dfrac{\partial w}{\partial y}\right)_x=0$, and (x, y, z) is a stationary point. The argument as it stands appears to depend on $\dfrac{\partial g}{\partial z}$ being non-zero, but since we may eliminate x, y, or z, it really depends on there being no point at which $\dfrac{\partial g}{\partial x}=\dfrac{\partial g}{\partial y}=\dfrac{\partial g}{\partial z}=0$; at such points (stationary points for the function g) there is no direction normal to the surface (e.g., the vertex of a cone). If such points occur, we do not consider them in investigating stationary points of f.

Equations (4.3) together with the equation $g(x, y, z)=0$ enable us to find the stationary points of the function f restricted to the surface $g(x, y, z)=0$. We now exemplify this procedure.

Example 4.2. Find the maximum values of xy^2z^3 subject to $x+y+z=1$.

We must solve $\dfrac{y^2z^3}{1}=\dfrac{2xyz^3}{1}=\dfrac{3xy^2z^2}{1}$, $x+y+z=1$. Suppose first that $y\neq0$, $z\neq0$. We then find $y=2x$, $z=3x$, so that $x=\frac{1}{6}$, $y=\frac{1}{3}$, $z=\frac{1}{2}$. To show that this is a maximum, let us put $x=\frac{1}{6}+h$, $y=\frac{1}{3}+k$, $z=\frac{1}{2}+l$, where $h+k+l=0$, since $x+y+z=1$. Then

$$xy^2z^3=\frac{1}{432}+\frac{1}{72}(h+k+l)+\frac{1}{48}k^2+\frac{1}{36}l^2+\frac{kl}{12}+\frac{lh}{12}+\frac{hk}{12}$$
$$+\text{terms of higher order in } h, k, l$$

$$=\frac{1}{432}+\frac{1}{144}(3k^2+4l^2+12kl-12k^2-24kl-12l^2),$$

putting $h=-k-l$,

$$=\frac{1}{432}-\frac{1}{144}(\ (3k+2l)^2+4l^2).$$

Thus $\frac{1}{432}$ is a maximum value. Of course, in the above argument we have simply been obtaining the Taylor expansion at $(\frac{1}{6}, \frac{1}{3}, \frac{1}{2})$; the vanishing of the linear term confirmed that we had found a stationary point. Clearly $\frac{1}{432}$ is only a local maximum; we may find points in the plane $x+y+z=1$ at which x, z are arbitrarily large and positive and y arbitrarily large and negative, so that xy^2z^3 takes arbitrarily large positive values.

Notice here that we 'neglected' terms of higher order in h, k, l; we are entitled to do this since we are only interested in the behaviour of xy^2z^3 for small changes in x, y, z and, as demonstrated in general in the proof of Theorem 4.6, this behaviour is determined by the second-order terms in h, k, l. The reader should beware of 'neglecting' terms without a theoretical justification for the procedure.

It remains to consider the lines $y=0$, $x+z=1$, and $z=0$, $x+y=1$. All points on these lines are stationary for xy^2z^3. It is clear that xy^2z^3 takes opposite signs at points in the plane $x+y+z=1$ arbitrarily close to any fixed but arbitrary point on the line $z=0$, $x+y=1$, so that no point of this line is a maximum or minimum for xy^2z^3. We examine the nature of the stationary points lying along the line $y=0$, $x+z=1$ by direct argument. If $x<0$, then $z>1$ and $xy^2z^3 \leqslant 0$ for x, y, z close to $(x, 0, 1-x)$. Thus all such points are maxima. If $x=0$, xy^2z^3 changes sign near $(0, 0, 1)$ which is therefore neither a maximum nor a minimum. If $0<x<1$ then $xy^2z^3 \geqslant 0$ for x, y, z close to $(x, 0, 1-x)$, and all such points are minima. If $x=1$, xy^2z^3 changes sign near $(1, 0, 0)$ which is therefore neither a maximum nor a minimum. If $x>1$ then

$xy^2z^3 \leqslant 0$ for x, y, z close to $(x, 0, 1-x)$ and all such points are maxima. It may be remarked that our analysis in this paragraph applies also to the function xy^2z^3 defined over the whole of three-dimensional space.

Example 4.3. Find the stationary points of the function $x^2+y^2+z^2-yz-zx-xy$ defined on the surface
$$x^2+y^2+z^2-2x+2y+6z+9=0.$$
We must solve $\dfrac{2x-y-z}{2x-2}=\dfrac{2y-z-x}{2y+2}=\dfrac{2z-x-y}{2z+6}$, $x^2+y^2+z^2-2x+2y+6z+9=0$. We demonstrate a technique of solving which is often applicable to this class of problem. Instead of attempting to solve the three equations directly, we put each ratio equal to λ (first removing the factor 2 from the denominator, for convenience). We thus get
$$2x-y-z=\lambda(x-1),$$
$$2y-z-x=\lambda(y+1),$$
$$2z-x-y=\lambda(z+3),$$
$$x^2+y^2+z^2-2x+2y+6z+9=0.$$

We now solve these four equations; adding the first three, we have $0=\lambda(x+y+z+3)$. If $\lambda=0$, we have immediately, from the first three equations, that $x=y=z$, whence $3x^2+6x+9=0$. Since this equation has no (real) roots, $\lambda\neq0$, so that $x+y+z+3=0$. Then $2y-z-x=3(y+1)$, so that $y=-1$ or $\lambda=3$; but $2x-y-z=3(x+1)$ and $3(x+1)\neq 3(x-1)$ so that $\lambda\neq3$, whence $y=-1$.

Thus $y=-1$, $x+z+2=0$, so that $x^2+1+x^2+4x+4-2x-2-12-6x+9=0$ or $2x^2-4x=0$, $x=0$ or 2. It follows that the stationary points are $(0, -1, -2)$ and $(2, -1, -4)$.

A word of warning should be given about these cases of 'stationary points with restraints', as they are sometimes called. There is an ambiguity about the phrase 'stationary points of $f(x, y, z)$ subject to $g(x, y, z)=0$'. We may mean 'those stationary points of $f(x, y, z)$ whose co-ordinates satisfy $g(x, y, z)=0$' or we may mean 'the stationary points of the function $f(x, y, z)$ defined on the surface $g(x, y, z)$

$=0'$. *Our meaning above is the latter*; any stationary point in the former sense is also a stationary point in the latter but not conversely.

Exercises

1. Generalize Theorem 4.1 to functions of three variables.

2. Generalize Theorem 4.2 to functions of three variables.

3. Prove that, in general, a function z satisfying $z_{xx} + z_{yy} = 0$ has no maxima or minima. If $z = \frac{1}{5} x^5 - 2x^3 + 25x + ax^3y^2 + bxy^4 + cxy^2$, find a, b, c so that $z_{xx} + z_{yy} = 0$ and then find the stationary points of the function.

4. Find† the maxima and minima of $\dfrac{x}{y} + \dfrac{y}{x} - \dfrac{(x-y)^2}{a^2}$.

5. Find the maxima and minima of $x^3 + y^3 - 3(x+y)$.

6. Find the maxima and minima of $\dfrac{x^2 + y^2 + 2x + 1}{x+y}$.

7. Find the maxima and minima of $(x-y)^2 + x^3 + y^3$.

8. Find the stationary points of the function $y^2 + 4z^2 - 4yz - 2zx - 2xy$ restricted to the surface $2x^2 + 3y^2 + 6z^2 = 1$.

9. Find the stationary points of the function $ax^2 + by^2 + cz^2$ restricted to the plane $x + y + z = 1$. Hence find the minimum value of the function $px^2 + qy^2 + rz^2 + 2yz + 2zx + 2xy$ restricted to the plane $x + y + z = 1$, where $p > 1$, $q > 1$, $r > 1$.

10. Find the stationary points of the function $x^2 + y^2 + z^2 - yz - zx - xy$ restricted to the surface $x^2 - y^2 + z^2 = 2$. Determine whether they are maxima, minima, or neither.

11. Show that if the range of the function $f(x,y,z)$ is restricted by the two conditions $g(x,y,z) = 0$, $h(x,y,z) = 0$, then the stationary points of f are given by the solutions of the equations

$$\frac{\partial f}{\partial x} + \lambda \frac{\partial g}{\partial x} + \mu \frac{\partial h}{\partial x} = 0, \ldots, \frac{\partial f}{\partial z} + \lambda \frac{\partial g}{\partial z} + \mu \frac{\partial h}{\partial z} = 0,$$
$$g = 0, \ h = 0.$$

(λ, μ are called the Lagrange multipliers.)

---

† This exercise is difficult!

CHAPTER FIVE

Appendix

In Chapter One we stated without proof the fundamental identity

$$f_{xy}(a, b) = f_{yx}(a, b)$$

(see (1.9)), and we have made heavy use of this identity in subsequent chapters. We did not prove it on its first appearance, not because the proof is inordinately difficult, but because we did not wish at that stage to hold up—or discourage —the reader who was concerned to understand the nature of the new concepts to which he was being introduced. In this brief appendix we supply the proof.

Let $f(x, y)$ be a function of the two variables x, y, possessing derivatives† f_{xy}, f_{yx} near the point (a, b).

Theorem 5.1. *If* f_{xy} *and* f_{yx} *are continuous, then* $f_{xy}(a, b) = f_{yx}(a, b)$.

We consider the expression

$$\Delta f = f(a+h, b+k) - f(a, b+k) - f(a+h, b) + f(a, b). \tag{5.1}$$

Then, if $G(y) = f(a+h, y) - f(a, y)$, we have
$$\begin{aligned}\Delta f = G(b+k) - G(b) &= kG'(b+\theta k), \ 0 < \theta < 1, \text{ by MVT,}\\ &= k(f_y(a+h, b+\theta k) - f_y(a, b+\theta k),\\ &= hkf_{xy}(a+\phi h, b+\theta k), \ \ 0 < \phi < 1, \text{ by}\\ &\qquad \text{MVT.}\end{aligned}$$

Thus $\dfrac{\Delta f}{hk} = f_{xy}(a+\phi h, b+\theta k)$ and so $\displaystyle\lim_{h,\,k\to 0}\dfrac{\Delta f}{hk} = f_{xy}(a, b)$,

by the continuity of f_{xy}.

† Recall that $f_{xy} = \dfrac{\partial^2 f}{\partial x \partial y} = \dfrac{\partial}{\partial x}\left(\dfrac{\partial f}{\partial y}\right)$.

The result now follows by the symmetry of $\dfrac{\Delta f}{hk}$ in the variables h, k; we must also have $\displaystyle\lim_{h,\,k\to 0}\dfrac{\Delta f}{kh}=f_{yx}(a, b)$, so that $f_{xy}(a, b)=f_{yx}(a, b)$. Actually we can do better than this; suppose we are simply told that f_{xy} is continuous and that f_x exists. We may then show that f_{yx} exists and that $f_{xy}=f_{yx}$, but we will not give the proof here.

To emphasize the non-triviality of Theorem 5.1, we give an example where $f_{xy}\neq f_{yx}$.

Example 5.1. Consider the function

$$f(x, y)=\frac{x^3+2y^3}{2x+y}, \quad \text{if } 2x+y\neq 0,$$
$$=0, \quad\quad\quad \text{if } 2x+y=0.$$

Then we verify that if $2x+y\neq 0$,

$f_x(x, y)=\dfrac{4x^3+3x^2y-4y^3}{(2x+y)^2}$ whence $f_x(0, y)=-4y$, provided $y\neq 0$.

On the other hand, $f_x(0, 0)=\displaystyle\lim_{x\to 0}\dfrac{f(x, 0)-f(0, 0)}{x}$
$$=\lim_{x\to 0}\frac{(x/2)^2-0=0}{x}=0.$$

Thus, in fact, $f_x(0, y)=-4y$ for all y, whence we easily deduce that $f_{yx}(0, 0)=-4$. Similarly, if $2x+y\neq 0$,

$f_y(x, y)=\dfrac{-x^3+12xy^2+4y^3}{(2x+y)^2}$, so that $f_y(x, 0)=\dfrac{-x}{4}$, provided $x\neq 0$.

But we may prove that $f_y(0, 0)=0$, so that $f_y(x, 0)=\dfrac{-x}{4}$, for all x, and

$$f_{xy}(0, 0)=-\tfrac{1}{4}\neq f_{yx}(0, 0).$$

Of course, neither f_{xy} nor f_{yx} is continuous at the origin, since there are points of $2x+y=0$ arbitrarily close to the origin, and neither is defined at such points.

Exercise

Show that Example 5.1 may be generalized to any function

$$f(x, y)=\frac{ax^3+by^3}{cx+dy}, \quad cx+dy \neq 0$$

$$f(x, y)=0, \quad \text{otherwise},$$

provided $bc^3 \neq ad^3$.

Prove that such functions f are discontinuous at the origin.

Answers to Exercises

Chapter I

1. $f_x = e^{x+y}(\sin(x-y) + \cos(x-y))$,
   $f_y = e^{x+y}(\sin(x-y) - \cos(x-y))$,
   $f_{xx} = 2e^{x+y}\cos(x-y)$,
   $f_{xy} = f_{yx} = 2e^{x+y}\sin(x-y)$,
   $f_{yy} = -2e^{x+y}\cos(x-y)$.

2. $f_x = \dfrac{x^2+yz}{x^2y} \log\dfrac{x^2z-y^2x+z^2y}{xyz} + \dfrac{x^2-yz}{x^2y}\dfrac{x^3z+y^2z-z^2y}{x^2z-y^2x+z^2y}$,

   $f_y = \dfrac{y^2-zx}{y^2z} \log\dfrac{x^2z-y^2x+z^2y}{xyz} - \dfrac{y^2+xz}{y^2z}\dfrac{x^2z+y^2x-z^2y}{x^2z-y^2x+z^2y}$,

   $f_z = -\dfrac{z^2+xy}{z^2x} \log\dfrac{x^2z-y^2x+z^2y}{xyz} + \dfrac{z^2+xy}{z^2x}\dfrac{x^2z+y^2x-z^2y}{x^2z-y^2x+z^2y}$.

3. $u_x = \dfrac{x}{\sqrt{(x^2+y^2)}} + 1$, $v_x = \dfrac{x}{\sqrt{(x^2+y^2)}} - 1$,

   $u_y = \dfrac{y}{\sqrt{(x^2+y^2)}}$, $v_y = \dfrac{y}{\sqrt{(x^2+y^2)}}$,

   $x = \dfrac{u-v}{2}, y = \sqrt{uv}$ $(r-\sqrt{uv})$, $x_u = \dfrac{1}{2}$,

   $x_v = -\dfrac{1}{2}$, $y_u = \dfrac{1}{2}\sqrt{\dfrac{v}{u}}$, $y_v = \dfrac{1}{2}\sqrt{\dfrac{u}{v}}$.

5. $1, 2$.

Chapter II

9. At most $p + q + rA \cot A$.

10. At most $\dfrac{pa^2 + qb^2 + (p+q)ab|\cos C| + rab\, C \sin C}{C^2}$.

11. $z = 2x + 4y - 5$,
   $\dfrac{x-1}{2} = \dfrac{y-2}{4} = \dfrac{z-5}{-1}$.

ANSWERS TO EXERCISES

Chapter III

1.　(i)　$-\dfrac{y+1}{x+1}$,　(ii)　$\dfrac{\cos y+y\sin x}{\cos x+x\sin y}$,

$\dfrac{1}{\log 1/3}$.

6.　$\dfrac{\partial(x,\,y)}{\partial(r,\,\theta)}=r$,　$\dfrac{\partial(r,\,\theta)}{\partial(x,\,y)}=\dfrac{1}{r}$.

7.　$-\dfrac{1}{2}\,e^{-2(x+y)}\left(\text{or}\,-\dfrac{1}{2(u^2+v^2)}\right)$

8. Definition:　$\dfrac{\partial(u,\,v,\,w)}{\partial(x,\,y,\,z)}=\begin{vmatrix} u_x & u_y & u_z \\ v_x & v_y & v_z \\ w_x & w_y & w_z \end{vmatrix}$.

9.　(i)　$r^2\sin\theta$,　(ii)　r,　(iii)　7.

11.　$\dfrac{a^2(x-x_0)}{x_0}=\dfrac{b^2(y-y_0)}{y_0}=\dfrac{c^2(z-z_0)}{z_0}$

Chapter IV

3. $a=-2$, $b=1$, $c=6$; the stationary points are the four points $(\pm 2,\,\pm 1)$.

4. The function is a maximum at $(b,\,b)$, $b^2>a^2$, value 2; a minimum at $(b,\,b)$, $b^2<a^2$, value 2.

5. The function is a maximum at $(-1,\,-1)$, value 4; minimum at $(1,\,1)$, value -4.

6. The function is a maximum at $(-1,\,0)$, value 0; minimum at $(0,\,1)$, value 2.

7. No maxima or minima.

8. The six points $\left(0,\,\pm\dfrac{1}{3},\,\mp\dfrac{1}{3}\right)$, $\left(\pm\dfrac{1}{2},\,\pm\dfrac{1}{3},\,\pm\dfrac{1}{6}\right)$, $\left(\pm\dfrac{1}{2},\,\mp\dfrac{1}{3},\,\mp\dfrac{1}{6}\right)$.

9. Stationary point is $\left(\dfrac{bc}{bc+ca+ab},\,\dfrac{ca}{bc+ca+ab},\,\dfrac{ab}{bc+ca+ab}\right)$;

$\dfrac{(p-1)\,(q-1)\,(r-1)}{(q-1)\,(r-1)+(r-1)\,(p-1)+(p-1)\,(q-1)}+1$.

10. The function is a minimum at the two points $(\pm\sqrt{2},\,\pm\sqrt{2},\,\pm\sqrt{2})$, value 0; $(\pm 1,\,0,\,\mp 1)$ are stationary points but neither maxima nor minima.

Index